Please leave your feedback, thoughts, and reviews on Amazon.

See Lieutenant Colonel McBrearty on CBS Los Angeles *Veterans' Voices*.

See Lieutenant Colonel McBrearty on CBS Los Angeles *Sports Central*.

Photo courtesy C.B.S., Los Angeles.

Follow John J. McBrearty's other literary works at:

www.JohnWritesHistory.com

FROM HOME TO HELL

Combat Journal 1

(Abridged Edition)

يوميات القتال

by

John J. McBrearty

Lieutenant Colonel, U.S. Army (Retired)

The views expressed in this publication are those of the author and do not necessarily reflect the official policy or position of the Department of Defense, the U.S. Army, the U.S. government, the States of California and Washington, and the National Guard.

Paperback ISBN: 978-1-966064-27-5

Hardcover ISBN: 978-1-966064-28-2

Library of Congress Control Number: 9781966064275

Cover Photo: 302nd Airlift Wing, Peterson Air Force Base, Co, May 18, 2026, photo by Staff Sergeant Amber Sorsek, USAF.

All uncredited photos in this book are the property of John J. McBrearty.

Photos credited to other photographers are their property and appear as a courtesy in this book.

All other maps, images, and graphics are from government sources and in the public domain.

This story is a work of fiction based on events that followed 9/11. This literary work's story, names, characters, and incidents are fictional. No identification with actual persons (living or deceased), places, buildings, or products is intended or inferred.

Published by Military Books & More Publishing (MBM Publishing).

Dedication

This book is written for those who will come after us—
so they may understand what was asked, what was endured,
and what was sacrificed during Operation Iraqi Freedom.

It is dedicated to the nearly one thousand soldiers with whom I deployed,
and especially to the five of our unit who never returned home.
They paid the highest price ever asked of an American.

These pages carry your memory, your service, and our shared legacy.

"In Shaa Allah"
("God willing" or "If God/Allah wills" in Arabic.)

John J. McBrearty
Operation Iraqi Freedom
4 May 2004

Twin Towers in New York, N.Y., Pre-9/11,
Photographs in the Carol M. Highsmith Archive, Courtesy Library of Congress,
Prints and Photographs Division.

"When chaos becomes routine, creativity becomes survival."

John J. McBrearty

Contents

Foreword

From Home to Hell: Combat Journal 1 (Abridged Edition) is an unfiltered, firsthand account of mobilization, training, and deployment during Operation Iraqi Freedom. Written by Lieutenant Colonel John J. McBrearty, U.S. Army (Retired), this abridged edition captures the psychological, emotional, and operational realities faced by citizen soldiers preparing for war.

More than a memoir, this book is a study in leadership, adaptability, and the quiet power of creativity under pressure.

John's writing feels deeply personal yet universally relevant. Whether discussing the war on terror, leadership in chaotic environments, or the cultural tides that shape our nation, McBrearty writes with clarity, humility, and wisdom.

This is a historical resource, a call to patriotism, and a master class in courage under pressure and deserves a place on every American bookshelf. A must-read for veterans, students, and anyone who values honor, country, and the lessons of history.

As a senior officer, Lieutenant Colonel McBrearty would have been welcomed to have command of one of my units. He is an extraordinary leader and a gifted writer.

Major General Peter J. Gravett
U.S. Army (Retired)
Author of *Battling While Black*

FROM HOME TO HELL

Combat Journal 1

(Abridged Edition)

يـومـيـات الـقـتـال

Preface

Written in war. Preserved in truth.

War is not glamorous. It is exhausting, bureaucratic, dangerous, and unforgiving. Yet within its harsh structure, something remarkable emerged: adaptation.

Creativity enabled soldiers to solve problems, lead, preserve morale, and retain their humanity. These pages are proof that adversity does not define us—our response to it does.

This is not a story about politics.

This is a story about people.

Citizen soldiers called from ordinary lives...

Into extraordinary circumstances.

When certainty vanished, creativity remained.

This is the *Combat Journal* series.

JM

NOTES FROM THE AUTHOR

1. Disclaimer: My sources used historical concepts, race, terminology, and references that are now sometimes deemed offensive, discriminatory, and politically incorrect. I in no way embrace such concepts or intend to use such terms as insults. However, we must preserve our historical roots and ensure that our documentation is as truthful as possible.

2. Although there may be similarities to real-world circumstances following 9/11/2001, all characters and units in this publication are fictitious. Any similarities to living people are strictly coincidental.

3. Our fictitious main character, Major Darby Hillcrest, the Armor Battalion's Executive Officer (BXO), narrates this journal—Lieutenant Colonel John J. McBrearty, USA (Retired), who authors the book series. Occasionally, the author will insert his current opinions, which will be identified as written in italics.

Introduction

Written in real time during a nearly two-year deployment, *Combat Journal* is a military memoir series drawn from a combat journal kept inside the war zone. These volumes document the lived experience of modern warfare—arrival, combat, leadership, moral conflict, and return—through the unfiltered voice of a serving officer. This series preserves memory, honors service, and records sacrifice for future generations.

From Home to Hell chronicles the earliest days of a National Guard battalion's deployment to Iraq—arrival, uncertainty, and the slow realization that nothing familiar applies anymore. These journal entries capture the psychological shift from civilian certainty to combat reality. This abridged edition introduces the raw, unpolished voice that defines the series and sets the foundation for what follows.

In this initial volume, Lieutenant Colonel John J. McBrearty, U.S. Army (Retired), takes readers on an emotional rollercoaster as his main character, Major Darby Hillcrest, mobilizes, trains, and certifies his battalion of citizen soldiers for combat in Iraq. Find out whether these part-time soldiers can meet America's expectations. This is a military memoir you won't soon forget.

Army Values

"LDRSHIP"

Loyalty

Duty

Respect

Selfless Service

Honor

Integrity

Personal Courage

The core of the Army's leadership centers on the seven Army Values. These values will be a central reference point throughout the *Combat Journal* book series.

Map by John J. McBrearty.
Source: © OpenStreetMap (data is available under the Open Database License).

Chapter 1

Warning Order and Mobilization

25 March 03

While eating lunch at a restaurant near our National Guard armory, I received a call from my sister, Sherry. She was visiting our parents at their nursing home, just outside of Philadelphia, Pennsylvania. Sherry stated she was with Mom and that it did not look good at all.

I insisted, *"Put Mom on the phone!"*

I told my mother I loved her and that Annette (my wife), Nina (my daughter), and Little Junior (my son) also loved her. I asked my sister if Mom had heard me, as I had not heard a response from her. My sister did not confirm one way or the other, so I insisted she let me talk to Mom again. Sherry put the phone up to my mother. I again told Mom how we all loved her. Then, my sister said Mom looked up at the stars and faded away. Mine was the last voice she heard on this sacred Earth.

That was a difficult situation. When my sister called me back to confirm Mom's death, I froze. I immediately fought back the tears, but later, I did cry... eventually. I think I was in shock. How do you react when your mom dies? I will never forget the feeling; it was as if I were saying, *"Wait a minute, I want to talk to her some more..."* but you can't. Time froze before my eyes.

At that point, while I was still at the café, my lunch had not yet been served. I threw twenty dollars on the counter and abruptly left. I returned to the armory. I had to arrange some critical paperwork, such as soldiers' promotions and stuff. Rule #1 in my field: NEVER get in the way of promotions and money for a soldier. Despite my problems, my soldiers always came first. I eventually filled out my leave papers and cleared my leave over the phone with the Old Man (a term of endearment for a commanding officer).

Family support briefings prior to mobilization.

25 May 03

1st Day

This was the day before Sunday Family Support Day, held in a large hangar at the former Norton Air Force Base in California. The day ended with the viewing of our family support coordinator's (Carmon Dwellers) heart-wrenching PowerPoint presentation, which she put together for our family support day (tomorrow). It tore me up.

The last picture was of a little boy hugging his father's leg. The father was dressed in Army fatigues and appeared to be leaving for war. I thought that was my leg in the fatigues and that was my boy, Little Junior, holding onto my leg. I fought to hold back the tears. I brushed it off and left the room, just in case. I never wanted my soldiers to see me as vulnerable.

25 May 03 (continued)

This was officially the worst day of my life. This is the day I lost my best friend, my mentor, my counselor, my coach, my rock, my everything. He was my father, Dr. Hillcrest, Ph.D. My father died peacefully in his sleep as he was battling the effects of acute diabetes. Although Dad fought a long, downhill battle with this debilitating disease, and his health rapidly declined in the last years of his life, you are never ready for that ultimate life-altering event when it happens: death. Over the years, I had gotten so close to my father; this loss was unfathomable. Moreover, I was the senior active-duty officer in charge of a National Guard armor battalion, preparing to deploy to war. I was working seven days a week for many months, and now this? Just sixty days ago, we lost our mom, and now, Dad?

What else does God have in store for me this year?

2nd Day

Fear of the unknown.

Will I come back from war in one piece? Did I tell my wife enough that I love her? My kids? I have not slept well at all this week. Some nights, I was up most of the night. I had some vodka to help with the process, but even that proved pointless.

Sometimes, the fear of the unknown, like combat, was so overwhelming that it was almost paralyzing. My poor wife, Annette, she, too, hasn't had a good night's sleep this week.

We stood down for the deployment after our initial call-up in March and again in May 2003. These call-ups seemed like dry runs for the real thing. My full-time staff worked many long hours preparing for the imminent overseas deployment. I literally worked seven days a week from March to November. My deployment started in March 2003 because I was never home from that point on.

27 Nov 03

It is Thanksgiving Day. Unfortunately, our family does not seem to be giving much thanks this year because of the loss of my parents. This was truly a nightmare. I was so close to my father, particularly in his last years. I called him several times a week. I loved him and respected him more than anyone. I will always miss him. I feel some comfort in knowing he is with me in spirit and by my side when I must make tough decisions. He even thought of writing me a letter from his heart in the event of his death. It was written in the early 1990s. It was a tough time in my life, both personally and professionally. Good old Dad wrote of how proud he was of me and that he would always be with me, even in death. It was such a personal letter; it is too bad my sister was the first to find it, open it, and read it, even though it was addressed to me.

Welcome to my world.

That is the kind of sister I have, always thinking of herself first. She has chosen not to support me in the war effort and, for some unknown reason, cut off all ties between our two families. Perhaps she is really against our nation's war on terror? I feel sorry for her as I think God is not a part of her life. If He were, I doubt she would have mistreated me so badly over the last few years. I do not hate her, although she has sometimes made me mad. I have turned our situation over to God. In the Bible, Jesus says, *"He who is without sin shall cast the first stone."* (John 8:7) I believe in that. I forgive her for her non-Christian ways, as she might not know what harm she has done to others.

I often pray for her and her family.

28 Nov 03

I went to work early, like I usually do. The unit will report in by noon; I arrived around 9 a.m. As I'd been preparing for a year-and-a-half military deployment, my office was a complete mess. I still had one more duffel bag to fill up. Master Sergeant (MS) Myler, a venerable senior noncommissioned officer (NCO) and Vietnam veteran, kept reminding me to pack my stuff so he and his crew could load it on the bonded carrier. He was invaluable. I always seemed to do my personal stuff last.

Preparing equipment for combat duty in Iraq.

29 Nov 03

This is the day.

Goodbye day. The day that families dread.

My wife and I both woke up early. She made lunch to take with me, very thoughtful, and I ate it later that day. We arrived at the National Guard armory at about 5:10 a.m. Most of the soldiers hadn't arrived just yet. Luckily, there wasn't an overabundance of families there for the send-off. As many were there the day before, I am glad, and it was very sad. It was sad to see soldiers with their families. You could see the young couples holding on to each other as lovers do. You could see older couples with families sharing moments as families do. There were an occasional tear and an occasional argumentative flare-up, but there wasn't much of that. It was sad to see all this.

When Annette and I said our goodbyes, it was like you see in the movies. It was surreal. I fought back the emotions and won the fight this time. My wife started crying. I kissed her and hugged her. I told her I loved her and would be back for Christmas if I could get a flight. I told her to take good care of the children, Nina and Junior. She cried, said the word "goodbye," and drove away. I waved and waved and waved. A Veterans of Foreign

Wars (VFW) type was at the entrance of the armory with what looked to be his son and family. Again, I could not show emotion in public. Boy, I felt sad. Oddly enough, my horrendous stomachache went away as Annette drove off. I think I had some psychosomatic things going on, knowing we were to say goodbye. Before Annette's departure, I was physically nauseated.

The family burden was lifted somewhat. I cannot explain it; I still loved my wife as much as I did ten minutes ago, when she was by my side, but now the weight on my shoulders seems lighter. This illustrates a soldier's coping mechanisms.

A combat deployment is harder on the families than on their soldiers.

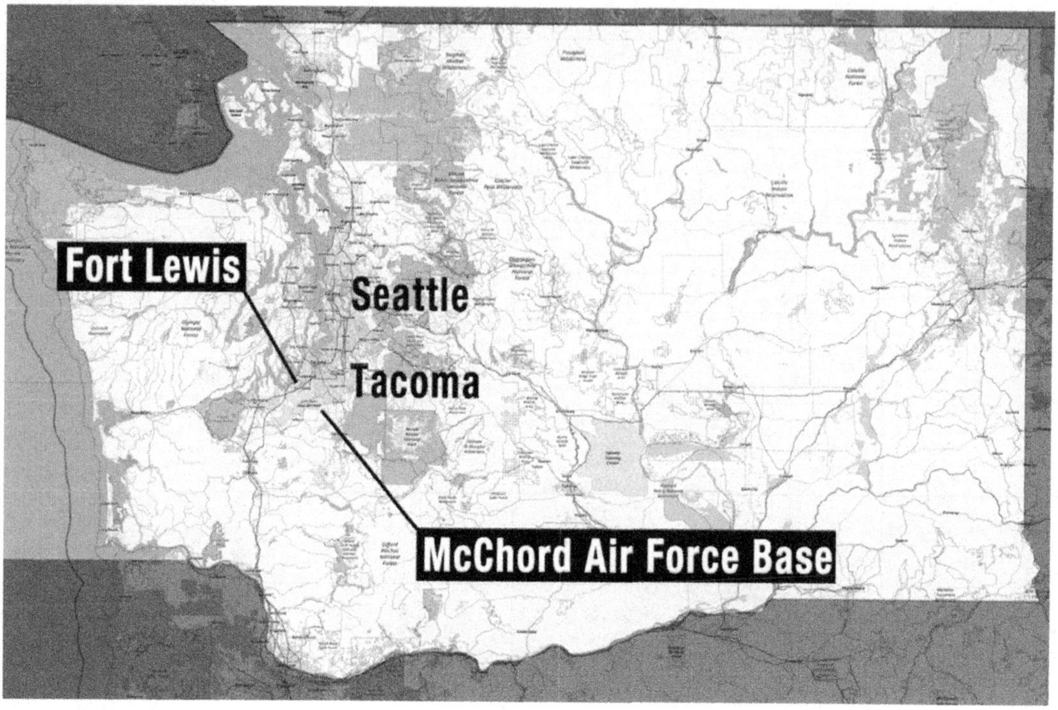

Washington State area map by John J. McBrearty.
Source: © OpenStreetMap (data is available under the Open Database License)

Chapter 2

Fort Lewis

30 Nov 03

We arrived via charter air yesterday, flying from Ontario Airport, CA, to McChord Air Force Base, WA.

We had a nice charter flight with South-worst Airlines; I mean, Southwest Airlines. Today, we conducted soldier readiness preparation (SRP), another hurdle to negotiate for this thing we call the love of God and country. Can I measure up? Will I finally get this far only to be bounced out because of my recent knee surgery?

As it turned out, my blood pressure was too high. I made sure to buddy up with the physician assistant (PA) during this phase of the journey. I worked out extensively back then and had relatively large arms. I informed the medic that my arms were too big to be used on the blood pressure machines and always needed to be tested by hand. The Army medic wasn't buying it. Luckily, our PA chimed in, and he bought it. In fact, the PA even had his own stethoscope handy, and I was in like Flynn!

Next hurdle: Anthrax injections. I was so concerned and apprehensive, I was borderline freaked out. I was surprised no one questioned my leg surgery (only seven weeks ago now), but I wasn't complaining. I worked hard during my physical therapy sessions and am glad I did. At least now, I can walk without a severe limp and without that terrible leg brace. Truth be told, I got the knee surgery in an effort to be physically qualified for our unit's deployment. I had meniscus issues and asked a surgeon, who, luckily for me, was also a veteran, to get me ready to deploy. I owe that doctor a lot and will never, *never* forget him.

I called home a few times and spoke with my wife. Several conversations went fine, but I sensed something was deep down wrong. I picked up on her sadness. I felt bad. That moved me so much that it inspired me to pick up my laptop and restart my journal. I feel helpless in not being able to help her. I know she is strong, and it must be intense for her to show hurt.

3 Dec 03

After morning chow, I walked out of the mess hall; it was early in the morning. We are situated so close to Puget Sound that I can sense the ocean and its grandeur just a few short miles away. To the south, you can see the forest as far as possible. A mist fills the air. It is solitude, the air so fresh. The sounds of wildlife are in the background.

This is the calm before the storm.

4 Dec 03

Today, we had classes about Iraqi culture, rules of engagement, field hygiene, public affairs, and the current threat in Iraq. I hope this information brings the reality of our situation into perspective for our soldiers.

I see some soldiers acting like they are attending an extended summer camp. Those are the ones who will create casualties for us. As the brigade commander, Brigadier General Killman, said last night, *"This historic moment marks the largest National Guard units mobilizing of this nature since the Second World War."* I think he got his wars wrong; he probably meant to say the Korean War. We will forgive him as he is not a native-born American.

Today was one of the worst days of my life (yes, unfortunately, this is a recurring theme for me). It was like seeing my life crumbling before my eyes. My wife called and sounded frantic. She was distraught. She informed me that my sister, Sherry, and her husband, Michael, were ordering their three children to cut ties with our family. Apparently, they received a letter from the attorney who is handling my affairs while I am on deployment. Those affairs included investments, taxes, legal matters, life insurance, inheritance, etc. Perhaps the Termagants were spooked by having an attorney in the mix? I will be gone for over a year, and I felt the need to have someone handle my financial and business affairs at home. It would not be fair to dump all that load onto Annette on top of raising Little Junior (2¼ yrs), and his sister, Nina (18). I chose to have a lawyer handle my affairs.

I don't know, but their behavior doesn't make sense. I have always loved and trusted my sister. I feel helpless because I cannot be there with Annette to help her through tough times like this. This experience has taken my mind off the ball. I have lost focus — this is unfair to the soldiers whose lives could depend on my decisions. Tomorrow is another day. I must measure up for my men, and I will.

War is hell, but it is worse for families.

God bless our soldiers. Good night.

5 Dec 03

Well, after one of the worst days in my life, today is a new day and was quite different. Today, I submitted four papers due for my master's degree class I am currently enrolled in. Despite the challenges of the mobilization, my parents' demise, the family send-off, etc., I somehow managed to get the work done. Today, I emailed the remaining papers to my professor. It was a good feeling. One paper was a thesis I am proud of. The paper is about the origins and history of the American Militia and the National Guard. Who knows, maybe someday, that research paper will be published.

I have issues with the Headquarters and Headquarters Company (HHC) Commander, CPT Ryker. He cannot get it together. He has no clue about his soldiers' accountability and training. I expressed my frustration regarding Ryker to the battalion commander (BC or CO).

For the past year and a half, while Ryker was company commander, I consistently offered him ideas and suggestions; he disregarded everything I recommended. This cannot happen now; the stakes are too high. LTC Slacker, our BC, agreed that he had to decide between keeping Ryker in the position and replacing him. I reiterated to him our obligation to bring our soldiers home in one piece and that, at times, these decisions are difficult but must be made.

I received a beautiful email from my father-in-law today. It mentioned my sister's reaction to receiving a letter from my lawyer. He also agreed that something was missing from the equation with Sherry. I responded to his email, stating that I could only pray for her as she was missing God in her life. I concluded this by observing the actions of Sherry and her family.

Annette told me that she had a long conversation with Sherry today. Annette explained to Sherry that life is too short to harbor such hatred and harsh words toward family or friends. She mentioned that one minute, she was saying goodbye to her uncle as he left on his vacation of a lifetime, his lifelong dream to drive across Canada and the USA. Well, that was the last time she saw him alive. She said you just do not know when it is your turn to go. She reminded her that I was on my way to combat, and there was no need for this hatred and harsh words.

6 Dec 03

Another day down.

Yep, it's only been a week, and we have not even gotten to Iraq yet, and I am counting the days. I love the soldier stuff, but the politics of pleasing the boss and the higher-ups gets to me sometimes.

RFI stands for rapid fielding initiative. Finally, the Army is getting it right. This is a program that front-loads combat troops with the latest "Force XXI" equipment. The Marine Corps and Army Special Forces developed this high-tech stuff. The equipment was specifically designed for such operatives.

Today, we received our initial issue with this equipment. This is a big morale booster for us. It makes us feel important because Big Army would not provide us with this high-speed equipment if it did not have ambitious plans for us. It helps us believe that we are part of "Big Army" (a slang term for the active-duty Army).

Later that day, most of the staff were in building 11D3 (or eleven Delta three, as we refer to it in military jargon), our battalion command post (CP). We were in the middle of moving offices around. Out of nowhere, LTC Slacker comes into our area and yells, "No fucken' way is the S4 (logistics officer) and all that minutia going to be outside my door. Move him in here!" You could hear a pin drop. All the staff looked at me. I bit my tongue because I wanted to laugh. I could barely hold it in. The staff belongs to me, and these issues should be addressed with me. I diplomatically told the colonel, "Sir, we will fire and adjust and make some movements of offices to accommodate your wishes." He left. I then grabbed the S1-Personnel Officer (CPT Randerson), the S3-Operations Officer (CPT p) Rickets (CPT p that means Captain promotable — it is an Army thing), and the S4-Logistics Officer (CPT Cantez). I said, "Let's establish our new courses of action (COA)." I spelled out five, then four, and then three COAs we all agreed upon as possibilities (this is part of the military decision-making process — an analytical approach to solving problems).

I then ran the three choices by the boss. He said, "Ask the sergeant major to come in." I wanted to say, "Sir, just make a friggen' decision already!" The command sergeant major (CSM) and I came in. I recommended the first COA, and he and the CSM agreed. So be it. In this COA, I relinquished my office to the staff. I felt they were the ones doing the work, and I wanted them to be comfortable so they could do their best. I also thought that if they saw a small sacrifice of mine for them here, it could pay dividends in the future.

At least, I hoped so.

Later that day, the chaplain and I rented a car at Fort Lewis's main post. This is sad; officers must rent cars at their own expense because the Army doesn't have enough vehicles to go around. We know we have a job to do, and we are willing to do whatever we need to do to get it done. We rented a Chevy Blazer for two weeks. We wanted a big enough vehicle to haul around troops. Often, you must run a soldier or two to the medical clinic, dental clinic, clothing issue point, training site, etc.

The USS Arizona (BB-39) was burning after the Japanese attack on Pearl Harbor on December 7, 1941. USS Arizona sunk at Pearl Harbor. Photographer: Unknown, Library of Congress Prints and Photographs Division

7 Dec 03

Today is the anniversary of the bombing of Pearl Harbor. If it were not for the local news, I wonder how many people of my generation and my daughter's generation would remember this date. I remember it from my father's story of where he was when he heard the news. He was attending a high school football game, probably with his brother, Edward. Like everyone else that day, I suspect my dad and his brother ran to the nearest enlistment center and signed up. They both entered the Army Air Corps. My dad was a machine gunner in B-17s and flew missions over Africa and Italy. He never spoke of his combat experiences, never ever. I heard from my cousins that he told my uncle some horrific things.

My father (left) and his brother (right) during World War II.

I guess I understand why my dad never spoke of his experiences; he was only eighteen. They say that generation never had a chance to live out their childhoods; they all became men/women overnight due to that war.

"The greatest generation!" General James Van Fleet and Tom Brokaw.

I wonder if I will have those same terrible combat experiences in the year that lies ahead.

We had battle staff training all day today. All primary staff of the battalions and brigades participated. The course of instruction will last five days or so. Lieutenant General (Retired) McYearson is the guest instructor in charge of training. He is the former 5th Army commander and a combat veteran multiple times. He is knowledgeable about the Enhanced Separate Brigades, the National Guard, and the Army.

In this classroom setting, some officers like to "grandstand." I do not go for this. I prefer to have my actions do the talking. However, some wish to impress others with their doctrinal knowledge.

There were no big emotional issues today. However, the BC and I spoke briefly about CPT Ryker, and the BC rolled his eyes, stating, "I have a problem there..." Thank goodness he recognizes this issue.

Later, I pulled my father's photo up on my computer. When I look at it, I get very sad. I cry inside. I almost cried outside, but I fought that off — I didn't want my soldiers to see me as vulnerable. I love him so much and miss him. The photo reminded me of the many times my father and I would work on projects around the house. We had quite a substantial workroom in the basement with every tool imaginable: table saw, jigsaw, router, vises, tool kits, etc. My father taught me how to operate all those tools, working on projects like the Cub Scouts Pinewood Derby cars. Those memories came to me as I looked at my father's photo. I miss that old gentleman so much.

9 Dec 03

Why write this journal?

I have several reasons.

First, if something happened to me, you know, the "it's going to happen to the other guy" stuff, I want my family, particularly my kids, to understand what was happening in my life then. Just in case it isn't the other guy, and it is my time to go, I want them to know what led up to the situation. I also want them to know a bit about their dad. I want my wife to know what was going on in my head and my life and how much I truly love her. I love her for not giving up on me when times were tough. She has always had faith in me, and I love her for that.

Today was a tough day, and it was a good example of things to come. We started early and waited for buses for an hour (the Army's "hurry up and wait" never seems to go away — you first see it in basic training, and you keep seeing it even twenty-four years later, regardless of your rank). Eventually, the bus arrived and drove us about thirty-five minutes into the country and to the training area. It was an LZ (landing zone for paratroopers). We did land navigation there. It was challenging for me as I am only two months post-op from knee surgery.

I tried to pamper my knee, being extremely careful not to lose my footing while traversing the rough terrain. Land navigation consists of navigating from point to point with nothing but a map, compass, and protractor. I have not done this since the USMC Officers' The Basic School (TBS) at Quantico, VA in 1982-83. That was some twenty-one years ago, and yes, I forgot a lot. When officers are in trouble, they often turn to a good NCO, which is what I did. I shared a map with Sergeant First Class (SFC) Barnum. He is a former infantry soldier and drill instructor. This guy knows his shit (this is a typical Army compliment).

He was quite the mentor. We had three hours to complete the course, which consisted of finding four points on the map. We finished the course in one and a half hours. That was good. Finding the points was tough; you had to navigate through the wilderness in the wet, cold Washington landscape for over four miles on foot. By the end, my feet were drenched, and my trousers soaked. The temperature was around 25 to 30 degrees Fahrenheit. From all of the physical exertion, I did not feel the cold other than in my feet.

My knee bothered me from this training. I iced it and tried to take it easy. The BC told me to stop limping so much, as it might draw attention to the issue.

I was stressed about this one. My knee was hurting.

I had gotten this far and refused to be dropped from the deployment until now. This is like being a high school varsity or college athlete who trains all year to win and finally makes it to the championship game, and it is questionable if they will play. Will I make it through this training?

Watching the news reports about the soldiers getting killed daily in Iraq is quite unnerving. I think this will drive our families nuts when we are away. Each time I see a broadcast, I pause. It hits me harder each time.

We get closer to the action each time and each day. Soon, it will be our turn in the shoot, as they say.

Back to today. We are on the second day of training with the MDMP. The MDMP process starts with conducting a mission analysis, so we focused on that initially.

The Military Decisionmaking Process

Key inputs	Steps	Key outputs
• Higher headquarters' plan or order or a new mission anticipated by the commander	Step 1: **Receipt of Mission**	• Commander's initial guidance • Initial allocation of time
		Warning order
• Commander's initial guidance • Higher headquarters' plan or order • Higher headquarters' knowledge and intelligence products • Knowledge products from other organizations • Army design methodology products	Step 2: **Mission Analysis**	• Problem statement • Mission statement • Initial commander's intent • Initial planning guidance • Initial CCIRs and EEFIs • Updated IPB and running estimates • Assumptions • Evaluation criteria for COAs
		Warning order
• Mission statement • Initial commander's intent, planning guidance, CCIRs, and EEFIs • Updated IPB and running estimates • Assumptions • Evaluation criteria for COAs	Step 3: **Course of Action (COA) Development**	• COA statements and sketches - Tentative task organization - Broad concept of operations • Revised planning guidance • Updated assumptions
• Updated running estimates • Revised planning guidance • COA statements and sketches • Updated assumptions	Step 4: **COA Analysis (War Game)**	• Refined COAs • Potential decision points • War-game results • Initial assessment measures • Updated assumptions
• Updated running estimates • Refined COAs • Evaluation criteria • War-game results • Updated assumptions	Step 5: **COA Comparison**	• Evaluated COAs • Recommended COAs • Updated running estimates • Updated assumptions
• Updated running estimates • Evaluated COAs • Recommended COAs • Updated assumptions	Step 6: **COA Approval**	• Commander approved COA and any modifications • Refined commander's intent, CCIRs, and EEFIs • Updated assumptions
		Warning order
• Commander approved COA and any modifications • Refined commander's intent, CCIRs, and EEFIs • Updated assumptions	Step 7: **Orders Production, Dissemination, and Transition**	• Approved operation plan or order • Subordinates understand the plan or order

CCIR commander's critical information requirement; COA course of action; EEFI essential element of friendly information; IPB intelligence preparation of the battlefield

Figure 9-1. The seven steps of the military decisionmaking process

5 May 2014 FM 6-0 9-3

Source U.S. Army Field Manual (FM) 6-0, 5 May 2014, in the public domain.

We raced back to the training area from the land navigation location to make our 1:00 p.m. start point (SP).

We made it, but the brigade did not.

We were to receive their warning order (WARNO), but did not actually get it until 7 p.m. (Hurry up and wait — again!) That is a joke, six hours late? I hope that doesn't happen in combat. We had to conduct staff estimates based on their lame warning order for a proposed mission in Iraq. This is our first glimpse of where we are going and what we might be doing.

It is fictitious for training purposes; however, it is a classified environment with security enforced. (This gets you thinking: Maybe it is not so fictitious?)

The day was spent having the staff spend time doing estimates in their functional areas in preparation for producing an operations order. It was a long day. We do not have adequate information from our higher headquarters to do our work properly. So, what does the American soldier do in these situations? Improvise.

It was a frustrating day. However, I saw some great things from our battalion staff. I saw mutual support from one another as they really started coming together as a team. This is good.

The BC was again frustrated; he went off a few times, which seems to be his leadership style, or lack thereof.

10 Dec 03

Today was another very long day.

Do we see a pattern here?

After a long day of battle staff training, we did night land navigation. The night training wasn't as hard as the day training for me because we had to do half of the walking we did in the day training. However, finding destinations at night with a map and compass is not an easy task.

Battle staff training is stressful for me because, as the battalion executive officer, I am the chief of staff. This is my show, and the microscope is on me. LTC Slacker chewed my ass pretty well today, too, and of course, in front of all the staff. I was seated next to CPT (P) Rickets. After the degrading ass-chewing, Rickets says to me under his breath, “Man, I must hand it to you, XO; no one takes ‘that’ as well as you. Morehead and I were discussing just that topic the other day.” Rickets was referring to taking an ass-chewing from LTC Slacker, and almost always in public. Rickets said he wanted to talk to me about it in private at a more convenient time. I think he is right. I hold up pretty well under that kind of pressure. In this instance, I knew something was adrift. Something wasn’t right about Slacker’s delivery. I think I put my finger on it. A colonel was visiting our classroom. I recognized him as Colonel Pesos-Madison. Yes, it’s a bizarre name, hyphenated and all. Well, I guess he got promoted or something because when I saw him less than a year ago, he was a lieutenant colonel. Anyway, he serves as the National Guard Bureau (NGB) Title 10 (federal active duty) liaison officer at the National Training Center, Fort Irwin, CA (NTC).

Why is it that when lieutenant colonels get promoted to colonel (full bird colonel), they become “dicks?” Then again, when I was a captain, I thought the same of majors. But now I am a major, and I never feel like a “dick.”

We have an upcoming NTC rotation in February. It is typical for units deploying overseas or into a theater of war to get a practice run at the NTC or similar facility. It is intense training against what the Army calls “the world-class OPFOR” (opposing force).

The training is unbearably tough and realistic. Ours will mirror our mission in Iraq. Additionally, this is how the active-duty Army validates your unit for combat duty. If you do not achieve the active-duty Army standards, your unit will get a “redo” and must conduct the training all over again until you get it right.

Anyway, I believe that Slacker was showing off in front of this officer. Strange, you bet! His ego often gets in the way. It could be the catalyst of his demise someday (but not under my watch).

More Slacker drama. Slacker was yelling because the S4 (CPT Cantez) wasn’t at the battle staff training, and he wanted him there. This is my position and current life; I know what Slacker wants, but I know what the battalion needs to be successful. If the battalion is successful, Slacker is successful. I knew Slacker wanted Cantez there, but Cantez had to take care of real-world logistical issues that day. If he didn't, we would be unable to conduct weapons qualification and other vital training, all of which are required for deployment. Slacker doesn't take the time to ask where Cantez is or why he is absent; he just flies off the handle as usual.

Welcome to my world!

Compounding the problem is the closure of Ft. Lewis for the entire post in a few days during the Christmas and New Year's holidays. There was no other time. I know this, and Cantez knows this, but Slacker is clueless. I am not one for making excuses. I just take the lumps because I know that, in this case, my actions are in the battalions' and Slacker's best interest. The proof is when we conduct our training successfully and meet our training requirements without significant incidents. If we had not addressed those logistical issues, we would not have been able to complete these tasks. Boy, Slacker would look bad if that happened, and so would the battalion. I look at it like it is my job not to let that failure ever happen, even if I must take some unwarranted ass-chewings along the way.

11 Dec 03

Today, we did more battle staff training. It went pretty well. However, most of the staff are sick, coughing up crud, sneezing, you name it.

In the morning, Chaplain Farrell and I went to get a cup of coffee at the local coffee shop on post. I bumped into the brigade command sergeant major (BDE CSM) while waiting in line. The BDE CSM is the highest-ranking enlisted soldier in the brigade. He is a cool dude. We discussed previous humorous land navigation experiences in our careers.

He mentioned that while at the Basic Noncommissioned Officer Course (BNOC), he observed several lieutenants conducting night land navigation, and many became lost, prompting a helicopter with a searchlight to be deployed to locate them. That is funny stuff. I told him that, while a 2nd lieutenant at Quantico, VA, I remembered that if we did not complete the land navigation courses in the standard curriculum, we would have to attend remedial classes on weekends. I remember driving off the post, heading home to see my love, Annette, and spotting lieutenants out with a map and compass, tramping through the forests of Quantico. I would say to myself, "Those sorry sons of bitches!" It was a crack-up.

Recently, there was news of three Stryker Brigade soldiers getting killed in Iraq. That brigade is from here at Ft. Lewis, so this is huge news. Again, every time we hear news like that, it hits home. I can only equate it to how those poor individuals must have felt before shipping out to Vietnam. They knew they were headed to a country where the Vietcong wanted them dead. And the war was not popular in the United States. I think we feel the same way. Although public sentiment is not as severe as during the Vietnam War, the situation is not substantially different. It makes you think.

Tonight, the chaplain and I took our rented Chevy Blazer and visited three of the company offices. Most soldiers were in good spirits.

I'll get back to you in approximately six months with an update on that subject.

12 Dec 03

Today, we finished up the MDMP battle staff training. It was quite a relief for me. We made incredible progress. So did our trainers from the 91st Brigade (training brigade—Active-Duty Army—the real deal). I believe we made great strides as a staff and team.

One of the most moving moments I experienced was hearing Lieutenant General McYearson (Retired) share his insights on the brigade's after-action review (AAR) of the staff training. He emphasized the importance of our staff planning, as lives will depend on our work. He stated that every time he hears of another American serviceman's death, it breaks his heart. He stammered as he was all choked up. You could see the sadness in his eyes as he meant what he said. It hit my heart. I feel the same way. When I watch the news, I brace myself for the possible news of another American's death. When I venture by a newsstand, I avoid the headlines. I just cannot bear to hear of more deaths in Iraq. It was moving to see this tough retired general be moved almost to tears at the mere mention of a soldier's death.

I held a staff meeting this afternoon at the battalion's command post (CP), building 11D3. Since we have been engaged in staff training and land navigation, we have not held meetings or worked outside the CP all week.

I staffed many upcoming events, including weapons qualification (across multiple weapon systems at multiple locations), concurrent training events, and the travel of a sizable force to the Yakima training center. We resolved many issues. Slacker was present and raised concerns about nearly everything.

Oy vey.

I do not necessarily like having Slacker present during my staff meetings because the staff work for me (directly), and I want control over them. His presence undermines my authority. If a commander attends the executive officers' staff meetings, it automatically becomes the "commander's meeting." Slacker does not understand this concept.

When the meeting concluded, I informed the staff that I wished to speak with them at my desk after the BC had left. Slacker said, "Who says I am leaving?" I think his ego is just out of control. He did not read that I wanted to deal with some staff issues alone. When he says things like that, he makes himself look like a jackass. I gathered the staff around my desk. I could tell Slacker wanted to know what was going on. I say, tough crap, this is my time with the staff. I told them we learned a lot at the battle staff training, and we must apply it to our real-world mission.

I mentioned that several made mistakes in their portions of the meeting, and when the BC asked for information, not all had answers. Some staff members even repeated the same mistakes. I told them they had their sections and needed to manage them accordingly. I also mentioned that they had NCOs and needed to establish a working relationship in which the NCOs supplied the staff officers with the required information to ensure their success. I also shared with them that I always used three-by-five cards when I was a staff officer. That indicates to the commander that you put some forethought into the staff meeting and provided a current status, e.g., personnel strength (numbers), vehicle maintenance status, etc. I hope they got the message. I concluded the meeting by thanking them all for doing a fantastic job.

I then took two officers aside and gave some mentorship guidance for their future meetings and briefings.

All in all, it was a good week and a good day, long hours but rewarding. I think the staff now has a greater appreciation for me and what a battalion executive officer knows and does. I can see it in their eyes. But I always ask myself, will we be ready for war?

13 Dec 03

It was a long day.

It is now 10:21 p.m., and I am finally in my rack, unwinding. There were numerous fires to extinguish today. It is a steamroller effect from our battle staff training. When you remove the leaders and planners from the equation, planning and coordination do not occur. With individual weapons qualification on the agenda for the next several days, substantial coordination was required, but it did not occur. This coordination included arranging for ammunition, fuel, food, and reserving ranges, as well as class VIII (medical supplies and combat lifesaver bags), among other items. Since I started typing, I have had two phone calls, and the S3 stopped by my room to discuss issues regarding tomorrow's training.

Again, welcome to my world!

I had a good staff meeting today at 5:00 p.m. I had all the staff present, along with the company executive officers and first sergeants. We de-conflicted many issues. The commander was also around (as he always seems to be for my staff meetings) and gave some comments and advice. This morning, I utilized some of what we learned in the battle staff training. I coordinated the staff's workload to cover us through Christmas. I created a comprehensive monthly calendar and recorded the key events, meetings, etc. I had all the staff give input into the process, which proved quite valuable. I set priorities for the staff to focus on and selected dates for required training like individual weapons qualification, make-up training, and suspenses —or deadlines for the battalion to meet.

14 Dec 03

A significant event occurred today.

We will remember this day for some time to come.

We got Saddam!

(Note: Iraq is eight hours ahead of the USA, and Saddam was actually captured on the 13th of December 2003.)

The U.S. Military Machine was given a mission, and it was successful.

With that part of the plan, anyway.

Saddam Hussein in 1996, before his capture. AP Archive - Public Domain.

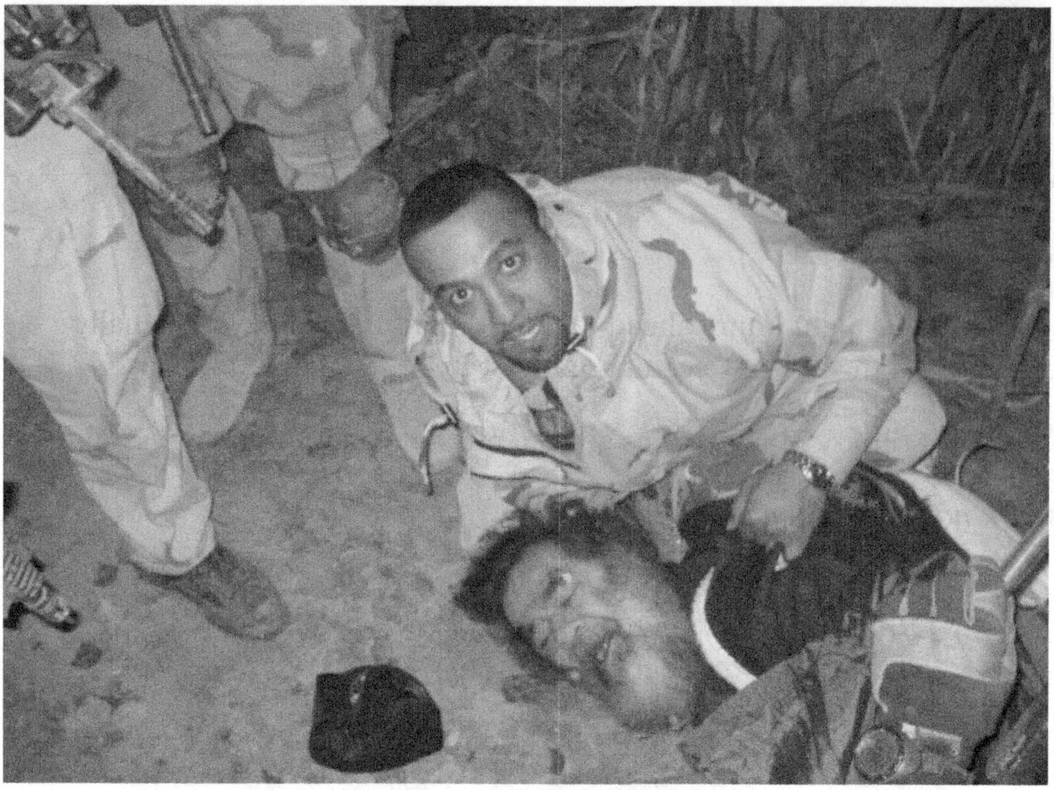

Saddam Hussein, captured by U.S. Army forces, is seen here in a "spider hole."
U.S. Army photo, public domain.

The local newspaper back home called me for a telephone interview. I told them our story and our reaction to the news. The newspaper ran a good article on us the next day; however, it was buried inside the paper again. As the old saying goes in the newspaper game, "If it bleeds, it leads."

So, we weren't the leading story this time in the paper because we had no bloodshed or death, but give it time, my friends. Give it time.

Last night, I bedded down (went to bed) at about 12:30 a.m. and awoke at 4:30 a.m.

I was on my way to the HHC area by 5:05 a.m. or so.

I drove a rental car and listened to the radio.

I was so thrilled to hear about the discovery and capture of Saddam Hussein, I could have screamed.

The adrenaline, the excitement, the sense of accomplishment! Those soldiers of the 4th Infantry Division who discovered him might never know how much we are enjoying and sharing in their victorious mission.

What a day!

Nothing could go wrong with a day that started like that. All the soldiers were overjoyed when they heard the news. They were pumped! The BC and CSM came into the battalion office beaming with pride and joy.

What a day!

Training also went well today.

I must hit the rack; good night, Saddam!

Peace.

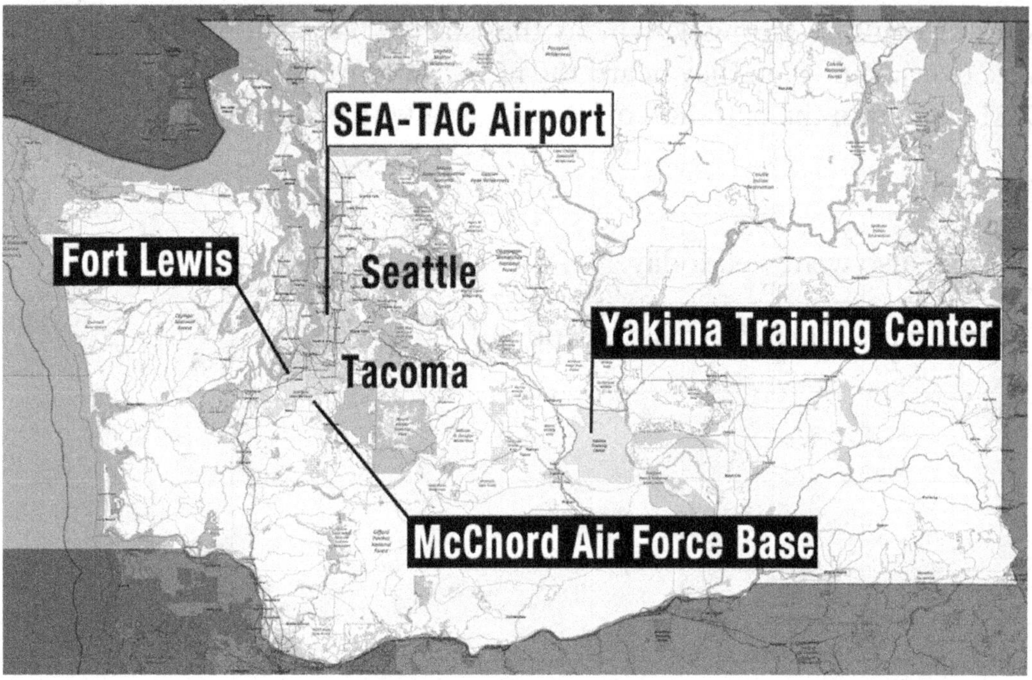

Washington State Area, map by John J. McBrearty. Source: © OpenStreetMap (Data is available under the Open Database License.)

Chapter 3

Yakima

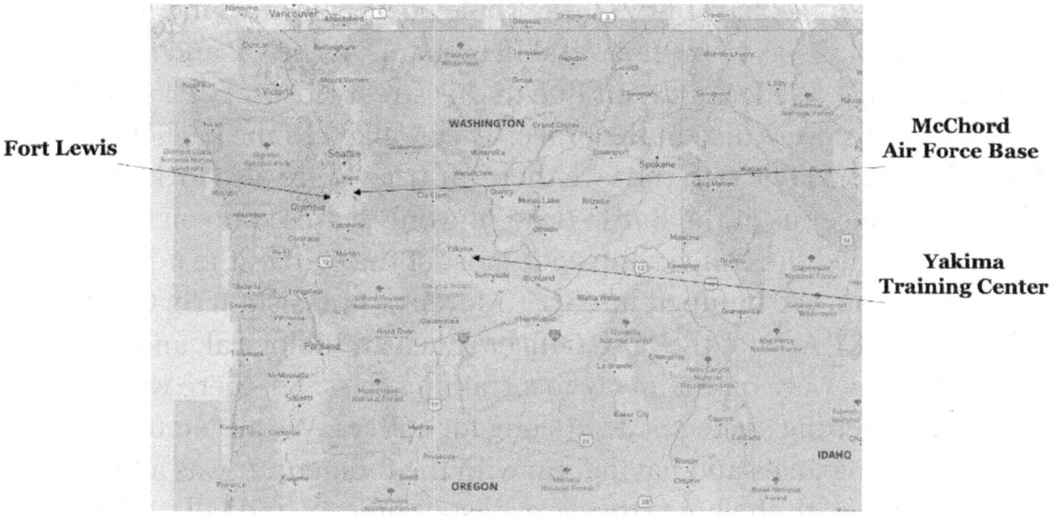

State of Washington

Washington State Area, map by John J. McBrearty. Source: © OpenStreetMap (Data is available under the Open Database License.)

15 Dec 03

Today, our main effort (priority of work) was our unit status report (USR). We conduct them quarterly in the reserve component (RC), but in the active component (AC) or active-duty Army, we conduct them monthly. It tells the Department of the Army our personnel, equipment, and readiness status. It made for another long day.

After a telephone call back home, I called it a night.

16 Dec 03

I have experienced a new level of exhaustion that I have not encountered in many years. I have hit the proverbial "wall." We did weapons qualification in the rain, and yes, it was also cold, particularly when the wind kicked up.

There is a moronic policy on the post I have fallen victim to several times already. Between 6:30 a.m. and 7:30 a.m. on Mon-Wed-Fri, the streets on North Fort (the northern part of Fort Lewis) are closed to traffic for physical training (PT). In other words, if you need to leave your parking lot to go to work, you are locked in until 7:30 a.m. I am sure this policy was implemented to reduce accidents or injuries. I think maybe some streets being blocked off is okay, but not the whole post. This morning, I was locked in again. I was to meet several lieutenants at the CP and take them out to the range. We lost an hour due to this street closure. CPT Arnold and CPT Cantez were also in the office. They walked to the bachelor officers' quarters (BOQ) to retrieve their car (they were also locked in). I told them to meet us at the Battle Bean, the only coffee shop on the North Fort. We invaded the coffee shop with most of the staff of the mighty armor battalion. We bumped into CPT Morehead (Assistant S3 Operations Officer) and 1LT Artera, the NBC Officer (nuclear, biological, and chemical); they were on their way to a training class. So, there we all were in our glory, getting some hot Joe (slang for coffee). We all "smoked and joked," cracking jokes and having fun with the moment. It was a bonding time. They are with the boss (they are all staff officers and fall under my supervision), and if the boss is playing, it is acceptable for us to play. We had some fun. After loading up with our "Joe" in hand, we drove to the M16 range.

We were wet, tired, and cold after the M16 (rifle) qualification. I shot well; I only missed one target out of forty.

We returned to the CP briefly. I bumped into CSM Fullerton (Division Command Sergeant Major). He is visiting us for a few days. Several other senior NCOs from our state were with him. I knew them from my time working at the division headquarters. I remember seeing one of the NCOs at the funeral of one of our soldiers who took their own life. The sergeant told me that he appreciated my taking the time to attend this soldier's funeral and that I was the only division staff officer present. He liked that. He said that if I ever needed anything in the way of favors, I should not hesitate to ask. Too bad I cannot recall his name. The lesson here is to always treat people with dignity and respect; you never know when your paths will cross again. Your honor and reputation will precede you. That means everything in this business.

Exhausted, I just crashed. I called Annette briefly. I also ordered her a dozen roses for our anniversary (tomorrow).

17 Dec 03

It's our twentieth wedding anniversary!

I called Annette early; I think I woke her up. I told her I loved her and to have a happy anniversary.

After waiting for what seemed like an eternity for the "all clear" from the morning formation street running Nazis, we bolted for the training area. We conducted improvised explosive device and unexploded ordnance (IED/UXO) training. Finally, we will gain hands-on experience we will use daily in Iraq.

We learned how to identify and maneuver around the explosives as well as how to extract a victim from one of those devices.

18 Dec 03

Now for our training events. After the morning breakfast and coffee, I rustled up the staff, and we headed to the HHC classroom. When we got there, a few soldiers were already in the room. We had to enter an isolated location with all the windows blocked and the doors locked, because the material we were working on was classified and tied to our mission in Iraq.

Today was the first beautiful day we've had here in Washington State. It has been only a month now. Clear and sunny, and in the fifties. I did not wear my jacket at all but did wear the woolly-pulley Army sweater (very comfortable).

The calm before the storm.

19 Dec 03

Today is my niece's (cousin #1's) birthday. I've been thinking about that for the last 24-36 hours. I was torn because I love her so much, but her mother is getting in the way of maintaining relationships with her kids. I am at a loss to understand my sister's peculiar actions and her apparent lack of patriotism in our war efforts. I emailed her a kind message, reminding her to let God into her life, for He will never let her down.

20 Dec 03

Today is Saturday.

I spent 3½ hours in the brigade executive officers' meeting. This is known in military circles as a "cluster fuck!" No apparent agenda?

Presently, our task organization is:

HHC AR

A Co., AR

C Co., AR

C Co., IN

B Co., EN

(NOTE: HHC-Headquarters and Headquarters Company, Co-Company, AR-Armor, IN-Infantry, EN-Engineers.)

Note: B CO, our armor company (Abrams tanks), is not there because it is under the brigade's task organization. Too bad; ours will be the first National Guard tanks involved in battle since the Korean War.

We currently have 625 soldiers in total, the biggest unit in the brigade. By the end of the deployment, those numbers rose to 1,000.

Later tonight, I walked out to the rental car parked in the parking lot and called Annette. She was depressed and was not too talkative. I turned the conversation toward my sister and her antics, which touched Annette's buttons. Annette took it hard when my sister's husband, Michael Termagants, yelled at her on the telephone regarding the fact that I had gotten a lawyer to handle my legal affairs while away. Funny thing, Michael would have done the same thing if in my shoes. But he would never be in my shoes because he is a coward. He could never be a soldier. I doubt he has ever even been in a fistfight in his life. A large man at 6'6" tall, he reminds me of what my Sa Bum Nim (master karate instructor 7th degree black belt) would tell us in my Taekwondo training: "The bigger the opponent, the bigger they can fall! The bigger the opponent, the more targets they present." There is a lot of truth to that in the martial arts community.

21 Dec 03

Sunday. I wish I could attend church, but it is not happening; I'm just too busy with work.

I went online and sent my wife, Annette, a heartwarming email. It reads as follows:

Hi, Annette,

I enjoyed talking to you yesterday.

I know you are down right now. I am frustrated, as I do not know how to help you. There are no quick fixes to our situation. All I know about life is that I always turn to the Lord when the going gets tough. I do not profess to be the most spiritually driven individual, but I think the Lord knows my true faith prevails when I am in a pinch. I believe faith is more than making a regular appearance at Sunday church services. It is in the heart. It is in one's character and one's day-to-day dealings with others and life. I cannot explain it; it is beyond my words. Faith and hope are what make us human, different from any other creature on the planet or in the universe.

I know talk is cheap. Our actions and choices are what make up our character and define who we are. I only hope and pray that this time for you will not be a bitter one. The mission at hand for our military is a humanitarian one. For once in our lives, we can take pride in the fact that "we" are doing something to help other human beings. Not just me, but Major Hillcrest's entire family is serving in the armed forces. I see it this way: we are not just serving the armed forces; in this mission, we are serving our fellow man/woman.

Certainly, 9/11 changed everything in our society. I feel privileged to make a difference. I truly feel honored to have the wife that I do. I believe that despite our differences and issues over the past two decades, our love for each other has truly prevailed. I feel God has rewarded us for this love and faith in him by giving us our son, Little Junior. What more is there to life than this?

I love you dearly. Please keep the faith.

I love you,

Darby

I met up with some of my staff officers at the chow hall. The brigade apparently took another weekend off, as we were the only soldiers in the mess hall. This is an absolute travesty, the imbalance of time off for us because we are from out of state. Not only can the brigade soldiers go home almost every night, but most of them also have the weekends off.

We get to work seven days a week, and *Slacker* gives us no *slack*. Wow, I think that I just came up with a good subtitle for this book.

"Slacker gives us no slack!" Just saying.

Later that night, Slacker knocked on my door. He asked me if I had issues with his leave policy (somebody must have tipped him off that I was researching the Army leave regulations — *oh, the loyalty?)* I said, certainly not, that I was only investigating the regulation, so it is all above board. I told him again that my only concern was that we were following Army regulations. If something happens to one of our soldiers while they are off duty, there could be liability issues. For example, if a soldier flies home for the holidays and does not have the proper leave paperwork on record, and if something happens to the soldier—death, accident, etc., their benefits for them or their family could be jeopardized. This is why the Army has regulations and puts everything in writing.

I also want to keep the boss out of any hot water that could arise. I do not think he understood my intentions (heck, Slacker, you cannot put a price tag on loyalty like mine!) I would never cross Slacker for anything! He is my commander, and I will follow him to my grave if I must; after all, I was trained by the finest fighting men/women on the planet, the United States Marine Corps. As I opened my journal with the seven Army Values, you will find Loyalty first on that list.

Anyway, this is what being a professional soldier is all about. The Marine Corps taught me this with some help from my father and a few good football coaches and karate instructors along the way. I also feel this way about former commanders. They are all surrogate fathers to me. I would never cross them for anything. I think they all know that, too.

Fear. I do not think of it often, probably because I am so busy. But with the forthcoming vacation home, I am sure I will have time to let fear appear. Before departing for our home state, I hardly slept at all. I even had to take Benadryl and some vodka to get sleep, but I did get it eventually. Unfortunately, I think my wife got even less sleep than I did. I think that is why she has come down with such a wicked cold; her body collapsed. I had many moments of fear about getting killed while in Iraq. Also, the anticipation of leaving was so stressful that I could not find the peace of mind to relax or sleep. I remember how Slacker and I disagreed about when we were to leave our state. I wanted to stretch it out over the Christmas holiday, and he wanted to get the show on the road as soon as possible. I just wanted more time at home, not only for me but for our soldiers.

Slacker thought the sooner we got the soldiers away from distractions, the better. Major General Jeffrey Ridley, the division's commanding general, had the final say; as a two-time combat veteran, he believed a final holiday at home would be crucial to our soldiers' morale. *Thank you!*

I will not have those problems again. After a month here, I have adjusted. The armor battalion is my new family for the next 16+ months. Not that my wife and kids are not important and first in my life, but my Army family will be all I have on my left and right flanks when facing the enemy.

22 Dec 03

PT from 6:00 a.m. to 8:00 a.m.

Christmas break went by like a flash.

Enough said.

Where did the time go?

It beats the hell out of me.

I will never forget being picked up at the airport by my wife, Annette. She drove up in her white Lexus 300, wearing a fashionable winter coat. Her hair was just the right tint of blond. She looked so hot! I felt fortunate to have her and to be able to provide her with some nice things in life.

I felt terrible on this trip, mainly about my situation with my sister.

The time at home was excellent. I just chilled. We didn't do anything extraordinary; we just spent time together.

29 Dec 03

Return to duty day.

I cried at home but hid it from my family. I wore sunglasses as we made our way to the airport. Small talk saved my ass from an emotional meltdown.

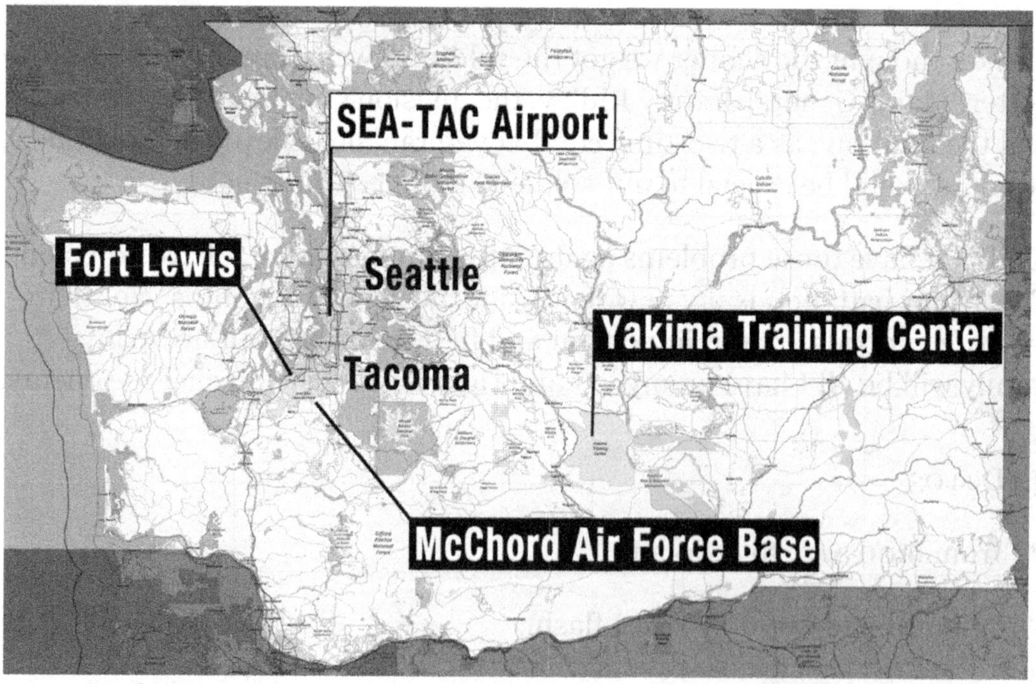

Washington State Area, map by John J. McBrearty. Source: © OpenStreetMap (Data is available under the Open Database License.)

I joined up with a group of NCOs at the Seattle-Tacoma (SEA-TAC) airport. We got a limousine ride back to the post. One sharp trooper whipped out a bottle of pricey tequila. The bottle was passed around the eight soldiers. We filled our glasses, and I raised mine, toasting a safe new year.

But would it be a safe new year in Iraq?

30 Dec 03

Back to the grind.

CPT Ryker has goofed things up again. The boss, Slacker, is fully aware of the situation. When we all took off on leave, Ryker left no one back to guard his weapons. Hundreds of weapons were not placed in a vault (which was unavailable) and were not guarded by an armed soldier, contrary to standard Army operational procedure (SOP). I asked the BC for his authority to launch a 15-6 (an investigation). This is a formal commander's inquiry into the facts surrounding the incident. There has been a pattern of a lack of accountability for the HHC weapons and personnel. I told Slacker, "No more, sir; we must do something." Slacker agreed. Later that evening, Slacker and I had a private discussion. He had already spoken with the brigade commander and planned to fire Ryker and possibly his first sergeant.

Now it is 10:00 p.m., quiet time, what I again like to refer to as...

The calm before the storm.

I will remember this night because the snow started falling to the ground as I chatted with my wife on the phone. I later retreated to my safe haven (my room), sat on the bed, and wrote in my journal. I popped in a *Rolling Stones* DVD my wife got me for Christmas. I will remember this night.

Tomorrow is New Year's Eve. Who knows how this year will turn out? Will all of us be here this time next year? My biggest fear in our whole experience is the possibility of not being there for my son. A boy needs his father. I am okay with the other aspects of my life. Nina is older now, and Annette is more than capable of starting a new life without me in the future — she is quite a looker. It is sad to talk about, but this is my reality. God bless my family and my soldiers.

Good night.

31 Dec 03

New Year's Eve. Yes, we worked. I had a long and fairly hectic day. The brigade chief of staff informed me that one of our tanks in training at Yakima, WA, rolled off the road and was stuck in a ditch. However, no one was hurt. The drama never stops.

To celebrate the New Year, I visited a fancy restaurant in Tacoma, WA, with several battalion staff officers. I later called home and had pleasant conversations with the family.

California map by John J. McBrearty.
Source: © OpenStreetMap (data is available under the Open Database License)

4-5-6 Jan 04

Next up: the Naval Postgraduate Institute, Monterey, CA. At around 3:30 a.m., we took a bus to the SEA-TAC airport.

I will never figure out why the Army has these outrageously early start times.

At around 6:40 a.m., we had a Southwest Airlines flight to San Jose, CA. From there, we took another hour-and-a-half bus ride to Monterey, CA.

We stayed at the Navy Lodge, which was quite nice, as the field-grade officers (majors, lieutenant colonels, and colonels) had their own rooms. From there, we took shuttle vans to the Navy Postgraduate Institute. At noon, we started our three-day course of instruction, which covered Iraqi culture, history, and the current political climate.

These three days were the most informative to date, and they were well-spent.

I also concluded that the brigade officers do not like us, specifically our battalion. This informal setting included wearing our civilian clothes to class, and after-hours were spent in the beautiful town of Monterey, CA. I experienced nothing short of snobbery and elitism by the officers of the other units within the brigade. I had to say something to Slacker at the airport, as we flew home. I knew he felt the same way about Washington officers.

He said I was correct in my observation.

He added that here in Washington, their brigade was the only game in town and ruled the roost — unlike us, where there is an entire division and a major support command within the state. Slacker said they were arrogant and thought they were better than us in every respect. I hate to say this, but Slacker is right on the mark for once.

We made it back to Washington. Treacherous is the word for the driving conditions, as snow and ice were everywhere. Washington State sucks!

The post is a dangerous mess. Training was even canceled for the past two days due to inclement weather.

7 Jan 04

Slacker and I discussed a possible replacement for CPT Ryker as HHC Company Commander: CPT Castellano, the Battalion Maintenance Officer (BMO). Although he was not doing a stellar job as the BMO, he had the tenacity needed to be a commander. He will do fine.

The rest of the day was spent in a training class for all the staff, both officers and enlisted members of the tactical operations center (TOC).

8 Jan 04

My next big purchase for the deployment is protective body armor. LTC Slacker, CSM Hunts, and several others have already ordered and received theirs, and they appear to be exceptionally high-speed. They purchased the same vests used by drug enforcement agents (DEA), which cost $800.00. Unfortunately, I will have to purchase this. There is too much press about National Guard units not being equipped with proper equipment like adequate body armor, and I do not want to take any chances. The oak leaf (major's rank) on my uniform is quite a desirable target for hostile adversaries.

I received a call from Mick, one of our local newspaper reporters. He was working on a story about the recent shooting down of an Army Black Hawk medevac helicopter. He asked if I had heard about the incident, how I felt about it, and about other soldiers' deaths in general.

I told him the truth. I told him my heart skips a beat whenever I hear of a soldier's death. I must catch my breath when I hear this type of news. The feelings are growing as we get closer to our time in-country. I said that soldiers are dying every day in Iraq. We are headed for Iraq, and each day, we get closer to danger. I went on to tell Mick that what keeps me focused on my job at hand is the importance of my family having me make it home in one piece.

Last night, I started writing letters to my wife, Annette, my daughter, Nina, and my son, Junior. These were to be opened only in the event of my death.

My father also wrote letters to my sister, Sherry, and me to be opened after his passing. They were written in 1994 (it is now 2004). The letter was Dad's final goodbye to me. It was as if he wrote it from the grave. He did not miss a beat; he was so organized. He really was squared away with his investments and insurance, and his communication with me was superb. And now this final note: God, how I love him!

Well, it is my turn, just in case.

I tried to put my thoughts on paper about my relationships with my family members. I tried to tell them how much they meant to me in my life in the event of my misfortune in Iraq. Before my departure overseas, I will mail these letters in sealed envelopes to my dearest friend, Luke, with the instructions for dissemination in the event of my demise. Luke was my bunkmate in the U.S. Marine Corps Officer Candidate School (OCS), also known as Bulldog, at Quantico, VA, during the summer of 1982. We have kept in touch ever since and have become great friends. I trust Luke with absolutely anything. A friend like Luke does not come along often in life, and I cherish our friendship. He is a quintessential officer and a gentleman.

Infantry squad training.

9 Jan 03

"Day of the Gremlins."

I call today "The Day of the Gremlins." Just about anything that could go wrong did go wrong. To make a long story short, the problems stem from our soldiers not being at the right training location at the right time with the proper equipment.

Another gremlin, I received the most upsetting email I have ever received. It was from my brother-in-law, Michael Termagants. I received an email that appeared to have been sent to me inadvertently by Michael's internet provider, stating that the previous email had been rescinded. The email was full of curse words, using my name in a profane manner. He commented on my attempts to maintain normal relationships with their three daughters despite the differences between Sherry and me, and he wouldn't allow that.

The incident shook me to the core, to say the least. I got a glimpse of my brother-in-law's innermost thoughts, which were demeaning and derogatory toward me. I was deeply hurt. In addition to the uncoordinated training events of the day, I was exasperated.

10 Jan 03

Today, we participated in additional Army battle command systems (ABCS) training. This is a computerized way of tracking the battlefield.

I worked late, up to 8:00 p.m. So, I ventured to the chow hall, and lo and behold, I sat with the only other person eating, LTC Kabat, the COS. We spoke about business over chow. I get the sense he just does not care much for our unit, or me, for that matter. He asked me about our governor, Arnold Schwarzenegger, and his anticipated visit. I think Arnold is the only good thing this brigade sees in our unit. The Washington brigade gets to rub noses with a celebrity. Whoop-de-doo! I felt like saying, "Who the hell cares about fucken' Arnold? Where are the parts for my trucks?" Yes, I think I have changed a bit and think in a good way. I sincerely care about one thing and one thing only: my soldiers!

12 Jan 04

Today was Monday, but I'd swear it was Sunday. We are so busy, we don't know what day it is. Oh, I stand corrected (military jargon for excuse me); we know when the weekends are here—that is the time when all the brigade mess halls are empty. Maybe it's because the brigade gets the weekends off, and we don't! This has been a significant source of contention for our battalion.

Today, we participated in a day of urban warfare training conducted by several colonels from the 5th Army at Fort Sam Houston, TX.

I spoke to my wife regarding the letter I wrote to my sister. I reviewed it with her one more time. She said it was well-written, and it was important for me to try to clear the air about my issues. The letter explained why I hired an attorney to manage my financial and business affairs while I was away in combat. It also addressed my displeasure about how she had interfered with her children's relationships with my family, particularly with my daughter, Nina. The whole event dealing with my sister has been too much to bear. I refuse to deal with it once I am in theater. My soldiers deserve an executive officer who is 100% in the game, and I won't give them anything less!

15 Jan 04

We started our command post exercise. It was slow to start; disjointed and unorganized. We are using technology that uses computerized battle tracking and imagery. The boss seemed happy today, and that is always a good thing.

Simulation Exercise (SIMEX).

16 Jan 04

SIMEX. Staff late. Staff not synced. Very sick. No time to write this week.

17 Jan 04

SIMEX. Staff on time. I'm still very sick and have no time to write.

18 Jan 04

SIMEX. Arrived early, 1st day shift.

The exercise ended around 11:00 a.m. or so.

We concluded our SIMEX, and as with any military operation, had an after-action review (AAR). For this event, the reviews were mixed. The bottom line is that, due to the training, we are all much better staff officers/NCOs.

19 Jan 04

Today is Martin Luther King Day, and the base is basically shut down. The entire brigade looks to be off work, except for, you guessed it, our armor battalion (the out-of-staters). I held a staff meeting at 9:30 a.m. and let them go if they weren't swamped with work. I mapped out the timelines for the staff to complete operations orders (OPORDs) for four upcoming events: maneuver training and tank gunnery at the Yakima Training Center (YTC) in the State of Washington, a training exercise here at Ft. Lewis, WA, and eventual movement to the National Training Center (NTC) Fort Irwin, CA.

20 Jan 04

Today was productive.

The computer Nazis (I.T. professionals) busted a lieutenant and a staff sergeant from our battalion, so I assigned an investigating officer, who compiled statements and fact findings. Investigations like this one are quite common on deployments.

Today, we also didn't see much of the brigade again. We heard they were getting a 3-day pass...again! It must be nice. All I can say is that I hope that we, the armor battalion, will get our time off after the NTC rotation at Ft. Irwin, CA. Or will we?

In the afternoon, I gathered the staff to conduct the Military Decision-Making Process (MDMP) for two of our four orders to be produced. The orders included relocating the entire battalion to YTC, conducting live-fire exercises there, relocating equipment and personnel to NTC, and ultimately moving personnel and equipment to Kuwait.

21 Jan 04

Today, Slacker drove the staff and me absolutely crazy. He stomped around, looking into each office, then he would say, "Where's so-n-so?" I would respond very calmly, "Sir, the staff's requirement this week is to have each section manned. For the rest of them, they needed some downtime." I have learned sometimes to shut Slacker out; otherwise, I would go insane.

The staff produced two more OPORDs today, which were completed on schedule. And not too soon—the company commanders are asking for planning information about our various training events.

22 Jan 04

Today, the CSM (Hunts) and I had a lengthy private conversation outside my office. The joke here is that there is absolutely no privacy in our lives, particularly in the battalion command post. So, if I had to talk to someone privately, we would put on our hats and step outside the door right next to my desk. The CSM and I took that walk today. We both had to vent our frustrations to each other about LTC Slacker. Slacker is on his high horse again, complaining about the staff not coming to work early enough and not having enough staff here to jump at his every whim.

The sergeant major said it is pretty bad when he yells at him. They are close, and Hunts is the highest-ranking enlisted soldier and doesn't deserve to be yelled at, with very rare exceptions. Slacker fails to realize that we are the only battalion that hasn't had a day off since we were mobilized, except for the leave during Christmas (they closed the post, and no training was conducted anywhere).

Slacker needs to get clued into his people, as his antics are creating a real problem. This is a leadership problem.

We held a brigade chief-of-staff meeting tonight with all the battalion executive officers; it lasted two hours. A lot was covered. I think I made some points with LTC Kabat, as I came prepared with information and pertinent questions. In the end, he threw us a curveball by asking how morale was. He looked at me, and I rolled my eyes. He said, "Is it that bad?" I just said, "Sir, we haven't had any time off, and it is taking a toll on our soldiers."

He asked why there was no time off. I couldn't say that Slacker wouldn't let us have downtime (career suicide), but rather, I said the training schedule didn't allow it.

I went on to say that time off after the NTC rotation in California is crucial to our soldiers' morale. (I was planting the seeds for that mission.)

23 Jan 04

The big issue currently is morale; it sucks. We are all burnt out and need a break. I think just taking a road trip to Seattle would work for my staff. At a command and staff meeting, LTC Slacker made his closing comments. He went on to say how he believes training is more important than getting a night or a day off here and there. I just disagree; you must care for the whole soldier, not just Army training, but their morale and families, too.

24 Jan 04

It is Saturday, but I'd swear it is Sunday. You lose track of time when you don't get a day off for months. We are getting used to it by now. Something tells me circumstances will not be changing any time soon.

Thank God, LTC Slacker finally replaced CPT Ryker as the HHC commander. He put CPT Castellano in the position. He isn't a great staff officer, but he has a strong personality and could succeed as a commander.

He isn't the brightest of the bunch, but that is okay. He has good NCOs to back him up and a full staff.

I thought of my dad last night and again today. I thought about how he must have felt going through the same issues I am experiencing as he trained and fought in World War II. I miss him so much. I hope I never shame my father, as I am his legacy. He deserves more than that. The importance of one's own legacy and mortality has grown with me over time.

25 Jan 04

Interestingly, the *USA Today* front-page article from 23-24-25 Jan 04 states, "Guard survey hints at exodus." This article discusses a National Guard survey indicating that many Guard soldiers plan to leave after their deployments. The reasoning behind these decisions concerns the demands placed on them regarding Operation Iraqi Freedom and the war in Afghanistan. It states that the Army has never before been stretched so thin that it had to rely heavily on the National Guard and the Army Reserve. It quotes the defense secretary as opposing increasing the size of the active-duty military and insists it isn't necessary. (WTF?) Whatever happened to President Ronald Reagan's philosophy of "peace through strength"? As I see it, if the reserve component makes up 50% of our fighting forces, that is clear evidence that we do not have enough active-duty troops. Hell, I am only a simple little major, and I figured that one out!

The military museum at Fort Lewis, WA, is said to be haunted.
Photo U.S. Army in the public domain.

I spent part of the morning with SSG Cactus, the BC's driver/bodyguard. SSG Cactus informed me that he once saw a documentary on cable television that identified the ten most haunted places in our country. This military history museum on the post was #1 on the list. What the heck! That place always gave me the creeps as I drove by, almost as if someone was peering out the windows at me, like in horror movies. Yikes! If I can muster up the nerve, I must visit this place. I told Cactus that when I worked at the division headquarters, it always felt creepy in that building. The building was an actual museum of the division's past military campaigns and had a very long history. There are many dead soldiers associated with the division's previous wars. This stuff is just creepy.

At work, I basically put out fires all afternoon.

26 Jan 04

Nightmare. Today was another nightmare.

Slacker. Boy, did he piss me off today. At 1:00 p.m., he called and said, "Hey, Hillcrest, Lieutenant Colonel Slacker here; where are you?" Why does he have to state his rank in these situations? "Lieutenant Colonel Slacker here." I know who he is and his rank. What the hell! I am a fucking major in the U.S. Army, and his right-hand man who lives to make him look good and successful, and he must check on my whereabouts like I am some private? I can't stand this type of micromanagement. He did this shit back in California when I was his full-time active guard reserve (AGR) officer in charge of the battalion, and he was the part-time commander. I told him I had been waiting at the front gate for an hour to get a pass for our rental car.

When I returned, Slacker said, "Hillcrest, if you are going to be out of the area for hours, you've got to check with me." Screw you, Mr. Slacker! I am pissed. I need to tell him that I refuse to comply with his ridiculous requirement and that he is just fucking wrong. But non-confrontation is more my style.

National Guard armor (tanks) training in Yakima, WA, for eventual combat duty in Iraq. This was the first time the National Guard deployed armor into combat since the Korean War.

"Did I tell you, Washington State sucks!" That was a quote from one of our PFCs that made me laugh. He summed it up nicely and said the

Washington National Guard members hate us troopers from out of state. We can't figure that one out, but maybe they are jealous of our weather, our lifestyles, or both?

Today, there are more news stories on TV and in newspapers about dead or missing soldiers. We dislike hearing these headlines, as they have a negative impact on morale.

This time, a soldier was lost in the Tigris River. He was a member of the Stryker Brigade from Ft. Lewis, WA. As I said to that reporter from home a few weeks ago, I can hardly stand hearing the news about dying soldiers in Iraq, as it tears a hole in my heart.

It also bothers me that it doesn't have much impact on American citizens unless, of course, it involves a family member or close friend. I hate seeing soldiers being expendable to the whims of politicians and generals.

We do our job well, fighting the enemy, that is. However, now we are undertaking another job that the military traditionally isn't trained to do: restoring civil order and restructuring a country's governing body. Our NTC rotation is a first-of-its-kind stability operations and support operations (SOSO). I hear they have designed this training specifically for us and the units that will follow. Yes, this is history in the making for the U.S. Army and indeed for the National Guard, and another good reason for writing this military memoir.

One final note: I almost fell off my chair this afternoon. I cornered CPT Rickets and LTC Slacker to address which staff were going to Yakima with our troops. Slacker said he wasn't going. I can't believe it; over 400 of his soldiers are going to a field training exercise where live ammunition is being used (this implies danger), and he isn't going to be with them? That is just fucking wrong. Every other battalion commander accompanied their battalions to the field, but Slacker chose not to. I told him the battalion needed command representation. He didn't say anything.

It upsets me that Slacker doesn't see a need for the troops to see him out in the field with them. That is where a commander belongs, next to the troops, leading from the front. Slacker doesn't get that. He dislikes going to the field, which is hindering the creation of a healthy command climate (how soldiers feel about their leaders). I am afraid this type of behavior will bite him in the ass sooner or later. Soldiers are smart, and word travels fast. They will figure him out and won't respect him. Troops think that if I must be out in Yakima, where it is cold and miserable, the commander should be out here with us.

Could this be a foreshadowing of things to come with Slacker?

I respect Slacker because he is my commander, but he is no leader! He thinks he is better than everyone else. Whatever happened to the Army Values — selfless service, personal courage, etc.?

Good luck.

And good night. (10:25 p.m.)

27 Jan 04

Where did the time go?

Flash, is the day over already?

There is so much to do, so little time.

I am glad Ryker is out , and Castellano is in, because I no longer have to deal with Castellano as the BMO, which he was a dismal failure at. He was a train wreck waiting to happen. WO1 Cliffs is the maintenance chief, and he walks on water; I am sure he will pull through for us. The chief is knowledgeable about maintenance, vehicles, and the system.

28 Jan 04

There are many fires to put out, including buses, billeting, chow, the advance party (ADVON), vehicles, ammunition, and weapons. Now is crunch time: today the ADVON departs for Yakima.

29 Jan 04

I have found my release, Linkin Park. I bought what I thought was a CD today, Linkin Park's *Live in Texas,* and it turned out to be even better than that: a DVD of their concert footage shot in Dallas, TX. They are my favorite band, unusual for a 43-year-old, maybe, but it is what it is, and that is what I am. Their music sends a positive message to our youth. They are rockers — alternative rockers, but clean-cut, don't curse, and the band members are reportedly Christian-based. And they rock! Often, listening to them brings me to tears. Their music conjures up so many memories for me, particularly the loss of my parents. I feel so empty without them, in particular, my father. Linkin Park conjures up memories with him, like when he took me to the driving range to hit some golf balls. Neither of us was a good golfer, but that didn't matter. What mattered was spending that one-on-one quality time with my dad. I will never forget that, and I hope to return from this nightmare someday to share those experiences with my son.

Good night. Back to Linkin Park.

I guess I just figured out why I am so emotional tonight and befriended my buddy, Mr. Smirnoff. At the SRP this morning, I bumped into a soldier while waiting in line to see the judge advocate general (JAG), which is the military's legal branch. This soldier was a young, borderline geeky type, small in stature; somehow, we started conversing. He said, "Are you guys going to Iraq?" I said yes. He smiled with a brotherly look in his eyes and responded, "I just came back from Baghdad." I asked him questions, such as what it was like and if he had any recommendations. I also said, "How long were you there?" He stated he had been there for six months. He then stopped, and his demeanor rapidly change . He got really serious all of a sudden. He said, "Don't become complacent... it will get you killed!" He continued, "Go there with it in your mind that you will get hit and hit numerous times. Expect it. Be ready for it, and don't let yourselves become complacent." He said they went by the thought: *"Drive 55 and stay alive!"* In other words, speed! Get to where you need to get to in a hurry! Stay in the middle of the roads, despite the lines and laws; the IEDs are on the sides of the road, and they will get you. This type of rude awakening affects you. Additionally, everyone appears to be wearing desert camouflage uniforms (DCUs). We are all going and going soon. This is reality. I suppose this has struck a nerve. I am scared.

I don't want my children to be fatherless, and I don't want my wife to be a widow.

My sister ceases to amaze me as she has ruined the relationships between our families. Sherry cut off all three daughters from corresponding with me via email, and they don't call or return calls either. I sent her a letter to explain the severity of my situation and asked her to get over her trivial crap. We both lost our parents this year, and we must deal with it. I wish she would seek some counseling. Here is the letter:

> ***From the desk of:***
>
> ***MAJOR Hillcrest***
>
> ***Hello Termagants,***
>
> *I hope all is well with your family and that you had a nice holiday season. Thank you all for the wonderful Christmas presents we received from your family this holiday season.*
>
> (NOTE: She sent our family nothing at all for Christmas.)
>
> *I wish to communicate the magnitude of my current situation. Simply, I am going to war. If you have been watching the news, you know soldiers are getting killed daily in Iraq. Currently, there are over five hundred American fatalities, not to mention the numerous casualties incurred by our soldiers. Our unit will be thrown directly into the middle of this action.*
>
> *As the executive officer of this battalion, I am second in command. I am responsible for the training, combat effectiveness, readiness, health, and welfare of over 700 soldiers. This, along with the fact that as a field grade officer, I make quite a target for Iraqi dissidents, displaced Kurds, vengeful Shi'as, murderous Sunnis, or any other opposition to the coalition forces. With all this on my plate, I am in no position to handle my personal affairs.*
>
> *The most prudent and responsible thing to do was to have a professional handle my affairs. I refused to dump all my issues on my loving wife, Annette. Annette has her hands full, raising Junior, maintaining her career, and managing our home.*

I informed Sherry that I was deeply hurt by her actions, cutting ties with our family, your only immediate family on the Hillcrest side.

(NOTE: What I don't understand is how Sherry could deal with me as a business partner rather than her brother and family member, but when her little brother has a business/legal representative for himself as he is shipped off to war, she falls apart.)

The communication issue — I will never stand between your relationship and that of my children and spouse. I will never understand if the same does not apply to you. Your children are the only ones whose conduct your conduct will hurt.

Sherry, I want you to know something. Your actions and the harsh words you have said to your family about my family and me hurt Nina very much. Your relationship and that of your family with Nina will never be the same again.

She loves you all dearly. It is really too bad, as she had nothing to do with anything between Sherry and me. Annette and I are mature adults who are at one with the Lord.

A famous Western proverb states, "To err is human, to forgive is divine." I have prayed many times for you, can forgive you for your hurtful actions, and can move on.

A verse from James states, "If we confess our sins, he is faithful and just to forgive us of our sins and to cleanse us from all unrighteousness."

Grief is very difficult for us humans to deal with; the Lord knows I have had many sleepless nights over our losses. I have spoken with a priest several times and might recommend it to you; it helped me cope. Grief counseling with a professional is always an option for all of us.

Michael, I got your email. It is pretty strong language. I will never understand your anger toward my family and me.

To sum things up, I will not be accessible after I leave for Iraq, and if I don't make it back alive, at least now you know a little bit about how I feel.

A friend told me something just before they died,

"...Don't let life pass you by..." Incubus.

There is a lot of wisdom in that saying.

I love you all,
Uncle Darby

MAJ Hillcrest
XO, Armor Battalion

I certainly hope I don't get hit and that I do come back in one piece, but at least I told my sister and her husband where I am coming from. Still, no emails or calls have been returned from her children, my beloved nieces. Her children, like mine, are victims of my sister's reign of hatred. I pray for them all.

Back to Linkin Park.

Good night and God bless.

I prayed many times tonight. I miss my parents and want all my soldiers to come back alive and healthy. That means more to me than life itself. Mom and Dad, I guess I am growing up. It has been said that a male doesn't truly become a man until his parents die. I think there is a lot of truth in that saying. God bless.

I don't want to sound unbrave, but I am scared. When I hear what is going on over there in Iraq from the guys who have been there, it hits home. I want to live up to the Army's expectations for my soldiers. I also hope to live up to their families' expectations and get them back in one piece, safe and sound. I pray for strength and judgment to succeed in this task.

Mom and Dad, please look over my shoulder. I'll need all the assistance I can get.

30 Jan 04

Friday, I'd swear it was Saturday.

I woke up at 5:30 a.m.

Because I had to drive the chaplain to the bus stop to go to today's training, we barely made it past the street Nazis and to chow.

No one is in the mess hall, which is not a good sign. I knew something was wrong. The issue was buses.

I returned to the BOQ and lay down for some sleep. Not half an hour later, the chaplain said there was a problem. The buses arrived for a 7:00 a.m. pickup, and the soldiers arrived at 9:00 a.m.

I immediately put on my fatigues and drove to the HHC area to extinguish the raging fire. I couldn't find the S4, which pissed me off. I am putting out his fires, and he is nowhere to be found. I visited the buses, and there were about seven or eight of them, but most of the bus drivers were seated on one bus. I approached that bus and received a warm welcome for a change. It felt good. Two of the bus drivers were middle-aged women, and they seemed to like me. I am not used to being flirted with. I wear glasses because I haven't opted for the vanity of LASIK, and I don't wear contacts.

Well, to be honest, I am not sure whether the two women hit on me or just built up my ego a bit. Hell, I don't know how to read women. They said, "A colonel, man, we got someone important here!" They did the token touching. As my wife says, this is the modus operandi for chicks who want to get their digs in on a dude. Man, we guys sure can be gullible. My wife was quite right. I must say, after seventy days of being with nothing but men and the Army, I enjoyed the moment. Sorry, Annette.

I ran around fixing the bus situation when the S3 and, eventually, the S4 showed up. They saw me directing traffic and wanted to help. I told them I wasn't happy with the miscommunications with the pickup times, and we looked like fucken' idiots to anyone outside of our organization. I wasn't happy.

By the time I made it back to the BTN CP (battalion command post), it was probably 9:30 a.m. already. Slacker saw me and called me into his office. Rickets was also there. Slacker closed the door. I thought, this is it, I am doomed, my career is over due to this fuck-up. I guess the staff tipped Slacker off that I was out there at the crack of dawn trying to fix the problem. I think he appreciated this, and I could see it in his eyes, but you will never get a "that-a-boy" from Slacker.

We did a staff AAR, as any professional would do after an operation or key event. We identified our problem areas. Fortunately, Rickets is quite humble and was the first to admit if he or his staff were wrong or missed a suspense. These are pretty unusual qualities in a man: noble, profound, and humble, marks of a great leader.

Later that evening.

Slacker confided in me with a secret he told me to keep between us. He said General Killman told him he would relieve the forward support battalion commander. Kabat is going to replace Rosales. I am glad to see Rosales go; he was a pompous ass who gave us the world's worst support last year at annual training. Slacker hates this dude, and so do I.

2 Feb 04

Writing this journal.

Writing this journal gives me a mechanism for coping with all the stress. Of course, as I have previously stated, I also want my family to know what was going on in my life and my head if I were to die in action. This is important to me. However, I am unsure how I would cope without the release that comes from my journal entries. At the very least, I am leaving behind historical documentation about our deployment for generations to come. I believe my graduate studies in American history have instilled in me a profound appreciation for the subject.

3 Feb 04

I am in a great mood; Chaplain Farrell has returned from the field. I missed my battle buddy. He visited the troops in Yakima, where they conducted various live-fire exercises.

Today, I learned LTC Café has officially taken over as the brigade chief of staff (COS). I reported to him in the morning. He was surprised and almost taken off guard. I said I reported to him and sought his guidance and expectations. He had no answers; I do not think he thought about it until I brought it up. None of the other BXOs gave him such courtesy.

I went on to tell him I wasn't a "grandstander" type like some of the other battalion XOs. I told him I expect my staff to work with his staff on issues, and that they should bring them to me if they cannot resolve them at their level. If I cannot fix the problem, I bring it up with him. Hence, I do not have much to add in meetings. I said there should not be much to put out in these meetings if we do our jobs; however, some like to grandstand. Café agreed.

Desert camouflage uniform (DCU) ceremony.

7 Feb 04

The DCU ceremony was held today at a large stadium in Seattle, and it was a cluster fuck for us armor battalion soldiers. The ceremony starts at 2:00

p.m., and the first bus of troops departs North Fort Lewis at 8:00 a.m. Yep, a very long day on our feet. Oh yes, the feet. We had the honor/privilege of wearing our desert boots for the first time today. What a time for this, on our feet for eight or nine hours, all wearing new, not broken-in boots.

The ceremony was well done, with the governor of Washington in attendance as the "keynote speaker." My only heartburn, I guess, was with the "long-winded" speakers. What the fuck, over? Over is military jargon for when a soldier ends their piece of dialogue. First, it was the governor, for at least twenty minutes. Next was the I Corps and Ft. Lewis commander, a three-star general. This dude went on for what seemed to be an eternity, at least half an hour. At this point, my feet, in the new desert tan boots, were aching, and we lost a few soldiers (they fainted) from the formation due to either hangovers, dizziness, or who knows what else. Typical politicians never give up a good opportunity to toot their own horn. I say politicians because, trust me, becoming a general officer involves a lot of politics.

Next was the Adjutant General of the Washington National Guard, a two-star general. I do not even know his name, nor do I care what the fuck his name is.

Man, am I getting bitter here or what?

8 Feb 04

One more day closer to getting the hell out of here!

Attitude? You bet! I went to church today, and a bishop gave Mass, which was spectacular for me. And a fifteen-year-old Asian girl sang; she was incredible!

B Co. IN and A Co. EN were last-minute additions to the brigade's mobilization. Both companies came from California at 100% personnel strength and helped the brigade fill out its manning requirements for the deployment.

Slacker went to Fort Irwin, CA, on the ADVON. I dig it when LTC Slacker is away because it leaves me as the HMFIC (head mother fucker in charge

— Army jargon). The unit seems to operate much more efficiently without the BC. Interesting.

Soldiers treat you differently when you are in charge, even if it's only for a week. I worked all afternoon, followed by the illustrious task of doing laundry, which puts things into perspective in case I ever get a big head.

I then went to the post exchange (PX) to buy a Magellan global positioning system (GPS). The one I purchased has built-in maps of the entire world, including Iraq. It wasn't cheap, $269.00. Add it to the list, approaching the $4,000.00 mark. All for the sake of God and country. This is nothing new; when the colonial militia was formed, part of the affiliation criteria included providing their own weapons, horses, and provisions. Here we are, some three and a half centuries later, doing the same thing all over again.

Kids are funny.

Today, my daughter, Nina, says of my quotes in yesterday's newspaper article, "Geez, Dad, that graduate school you did make you smarter; you sounded intelligent in the article." It is funny that I think there was a compliment there; somewhere, like before grad school, I wasn't so bright? She said that before my graduate studies, I never talked that way. God bless her. When I told Annette this story, she found it amusing.

Other fires to put out included moving and loading our equipment's Conex boxes (shipping containers) and initial planning for our next USR, which is due in three days. It never lets up.

MG Ronroe, the adjutant general of our state, is arriving tomorrow, and I have to spend Tuesday with him. Big whoop!

I am starting to get emotional thinking of my parents. I miss them. Church attendance brings this out, and I think I am a better man for it. It conjures up memories of my father and me going to Mass together. It is funny; my friends, growing up, told me their parents made them attend church on Sundays. They pointed out that I was the exception to that rule because I always asked my parents if I could attend church every Sunday. Hopefully, I am still a good Christian. God, please look after my soldiers.

Money. It has no meaning for me right now. Facing combat, injury, or death of my soldiers, and even my own questionable mortality, when it comes to money, well, forgive me for my "French," but I could wipe my ass with it for all I care. I have tried to stop spending money on this deployment, but the government doesn't always come through for us. I spent nearly $300 on that GPS today because I never want to risk being lost while in the military and in charge of soldiers. I look at life simply now; if I like it or need it for a mission, I buy it.

If I join my parents in Heaven this coming year, I want everyone to know I was pretty satisfied here in my mortal life. I also have the peace of mind that my family will be taken care of; between my serviceman's group life insurance (SGLI), my additional maxed-out life insurance policies I took out just before deployment, my military pension, and my thrift-savings retirement plan, my family will be financially set for multiple generations to come.

I am a profound optimist and figure if good old Dad made it through WWII, I could make it through this Iraq nightmare. I hope!

9 Feb 04

You do not even want to know about my day today. I got a call from the XO of ENG Co. at the tips-n-scales informing me that a cluster fuck was in progress with our unit. I tried to call both the unit movement officer and the battalion commander, but I received no answer from either.

Armor battalion soldiers are preparing to load equipment on a cargo ship docked at the port of Tacoma, bound for Iraq.

I then went to the ENG XO's BOQ room; he answered the door while dressing. The poor bastard was putting out fires until 2:30 a.m. last evening for the advance party. Our equipment was too heavy for the plane's lift, and we had to leave several thousand pounds of gear on the runway. He assisted in recovering that gear.

We drove to the tips-and-scales and met with the civilian officer in charge, a tall, retired Air Force first sergeant. He was a dick! We appeased him, kissed his ass, and tried to smooth things over. Our vehicles were not on time at the scales, and some were not stenciled properly. The problems were minor, but the previous day's events at the scales were disastrous for the support battalion (SPT BTN), and we are now feeling the aftereffects.

This is a nightmare: I must have aged months from this day. I received a call from the COS LTC Café, asking me what the fuck was going on. Man, the stress. I noticed a gathering of unfamiliar soldiers around a large white van, which is typical of a government service automobile (GSA). The adjutant general of California, MG Ronroe, and the director of the Army National Guard of California, BG Roams, were here to see us. I thought, what the fuck? No one told us they were coming. The timing was jacked. I met with BG Roams. I introduced myself: "Sir, I am Major Hillcrest, the executive officer of the armor battalion; our commander, LTC Slacker, won the toss of the coin for the advance party." He laughed and said hello as if he knew me. However, we had never been formally introduced until now.

MG Ronroe walked up to the support soldiers from our state, who were there on an annual training status (two weeks — big f'ing deal! Try eighteen months!) Oddly, he wasn't approaching the soldiers in our armor battalion and never said hello to me. I was offended on behalf of our soldiers and our unit. I refused to run up to him and say hello or to conduct small talk. Fuck this out-of-touch prick! Hillcrest does not kiss ass!

I was perturbed that MG Ronroe blew me off, blew off our soldiers, and blew off our unit (I guess that makes him a blowhard? Just saying.) I was more insulted that he blew me off later at dinner. Not even a hello or conversation. It was the most boring night of my forty-three years on this planet, other than talking to LTC Wieffer, his sidekick. I felt insulted again for our unit, our soldiers, LTC Slacker, and ultimately, myself. The conversation was never about the armor battalion and its deployment (excuse me; this is why all these Office of the Adjutant General pukes are even here visiting). I just do not get it.

I raised the topic of a three-day pass for our troops after our Fort Irwin training. This was a hard one for BG Roams to swallow. I confirmed with him that this was what BG Killman decided and agreed to. I made this decision because no one else would. Slacker wasn't in the staff meetings with the brigade, and I was, so I made the decision for our troops' sake. Someone had to step up to the plate, and I did. Whatever happens, happens. What can they do to me? Dress me up in desert fatigues and send my ass to Iraq to get shot at?

I am bitter as shit!

I care dearly about my Armor Battalion family so much that words cannot describe it. I guess I learned a hard lesson; we only have each other, the Armor Battalion. I love my men and will do anything for them.

What a fucked day; I could not even take a dump without getting a phone call. Slacker called to say hello. I called him back several times but was unable to get through. Eventually, he called me back. I told him things were getting done and not to worry. He sounded relaxed, and I am glad of that. That dude needed a chill pill. It is amazing what a loving woman's company can do for a man. I look forward to my turn. I love you, Annette.

10 Feb 04

LTC Weiffer told me last night that what he remembered from his mobilization (post-9/11; he was one of two National Guard infantry battalions from our state called up to immediate duty for a year of homeland defense missions) was how fast things moved and how much he demanded of himself. I concurred, as only someone who has been in his shoes could. We agreed that we push ourselves to an unbelievable limit: Push, push, and push some more. For example, tonight I am working until 9:00 p.m., which is not unusual, and I wake up the following day at 6:00 a.m. This type of schedule has gone on for months now, seven days a week.

I had an excellent day accompanying the TAG and conversing with BG Roams. We visited the reflex fire range and observed the soldiers conducting live-fire and maneuver training.

Leschi Town at Ft. Lewis, WA, is part of a combat village that resembles the Middle East.

We then went to Leschi Town and toured the facility. It is a mock-up of an urban environment used for training purposes. We watched soldiers conduct dry runs and rehearsals for training in Leschi Town. The training was Hooah! B Co. IN was the company recently deployed to backfill for the brigade. This unit deployed to Southwest Asia (SWA) two years ago, when I was the deputy G1 personnel officer, and conducted their SRP. This unit will do well as they are infantry and have already been to the Middle East (they have been there and done that). Even in the reserve component, infantry troops are what we call in the Army: "Hooah!" They are gung-ho and tough customers. Like the Marines, they are trained to close in and kill the enemy, which they do well.

11 Feb 04

Dinner at BG Roam's house with LTC Weiffer. They stayed in the very important people's (VIP) quarters on the post. In attendance from our battalion included company commanders, 1st SGTs, CSM, S3, and me.

MSG Studebaker from the MP company was there to give us a real-world account of life as a soldier in Iraq. This was scary stuff and a real wake-up call. We had a lot of one-on-one time with the general last night, which is quite unusual. They asked me a lot about BG Killman; I told them I respected him and that he was a great commander. (No matter how awful they might treat me, I always plug my bosses when given the opportunity — it is called playing the game.)

Again, hearing what that MSG had to say about his recent experiences in Iraq was a true come-to-Jesus meeting for all of us. The MSG saw death and destruction up close and personal.

Soon, it will be our turn.

I wonder how I will react under enemy fire.

12 Feb 04

Hectic is an understatement for today.

The brigade executive officer's meeting started at 4:00 p.m., lasted two hours, and included final details of the movement off post. The COS took at least ten calls during the meeting. Family support staff called me about issues with the dates for our family support-coordinated send-off. Oh, yes, now for the big event: one of our own soldiers was arrested for treason. A specialist from the other Armor Battalion, a Washington National Guard soldier, was arrested for being a member of Al Qaeda and giving information to the enemy.

What the hell?

This story hit all the news agencies. In our staff meeting, they addressed the official answer: "The Specialist Yadayada's case is currently under investigation, and we cannot comment until the investigation is closed." I held command and staff at 6:00 p.m. and addressed the issue of the arrested brigade member.

13 Feb 04

This is it, the last day on this God-forsaken post and state. Yes, we hate this fucken' place and this fucken' state. You can only imagine how fast-paced and hectic today was, as we packed, loaded, and prepared approximately 700 soldiers to mount up.

LTC Weiffer called me a bit later and gave me some details about our proposed three-day pass, and I almost jumped out of my skin with excitement. No shit, I danced and sang out loud. I was so happy. My soldiers must have thought that I was nuts. Weiffer said that BG Roams and BG Killman received authorization from I Corps, and the 5th Army approved our three-day pass and family day following the Fort Irwin training. He mapped out the days, including one additional day beyond what we requested. I ventured to see BG Killman to verify the plan. He said yes, that this plan was approved and fully endorsed. I was so happy. I hate to say it, but I think if Slacker had been involved with the process, this pass would never have happened. He doesn't like to make waves, and making this pass happen, well, it was a friggen' tidal wave.

A call from a family support volunteer woke me up. (I literally must have gone to bed with the phone in my ear, talking to family support folks about our three-day pass.) From there, I woke up the chaplain, and we hit, you guessed it, Starbucks. Today, I splurged and got a cinnamon biscuit with my Venti house coffee. Between waking up and my first cup of coffee, I had no fewer than eight to ten telephone calls.

Los Angeles / Victorville / Barstow area map by John J. McBrearty. Source: © OpenStreetMap (data is available under the Open Database License).

Chapter 4

Fort Irwin

14-15-16 Feb 04

We flew to the former George Air Force Base (AFB), which is in the high desert of Southern California. George AFB was officially decommissioned (shut down) in 1988 as part of the Base Realignment and Closure (BRAC) process at the end of the Cold War. It is currently called the Southern California Logistics Airport and is informally known as George Air Logistics Base. We then took buses from there to the National Training Center, Fort Irwin, CA (NTC).

I saw my family at the PX. Annette looked good, and Junior was talking like I had never heard before. He was seated in a little car, which was actually a modified shopping cart. He was talking about wanting the car. Man, how his vocabulary has grown. I finally got to spend some good quality time with the family.

16 Feb 04

Up early, 4:45 a.m.

After visiting with the family, Annette drove me to NTC.

17 Feb 04

I attended the brigade's command and staff meeting for the BC because our commander is, again, missing in action (MIA).

Home sweet home while at Fort Irwin, CA. (Yes, this is an actual circus tent!)

18 Feb 04

At NTC, we conducted reception, staging, onward movement, and integration (RSOI) operations in the rotational unit bivouac area (RUBA).

I could not write during this mobilization phase because we were in the field (NTC-hell!) Here is the situation: The entire brigade lives in five or six colossal circus tents. Our battalion of almost one thousand soldiers lived together in one of these big tents. All the cots and sleeping bags were lined up in rows according to company, squad, etc. (Ducks in a row.)

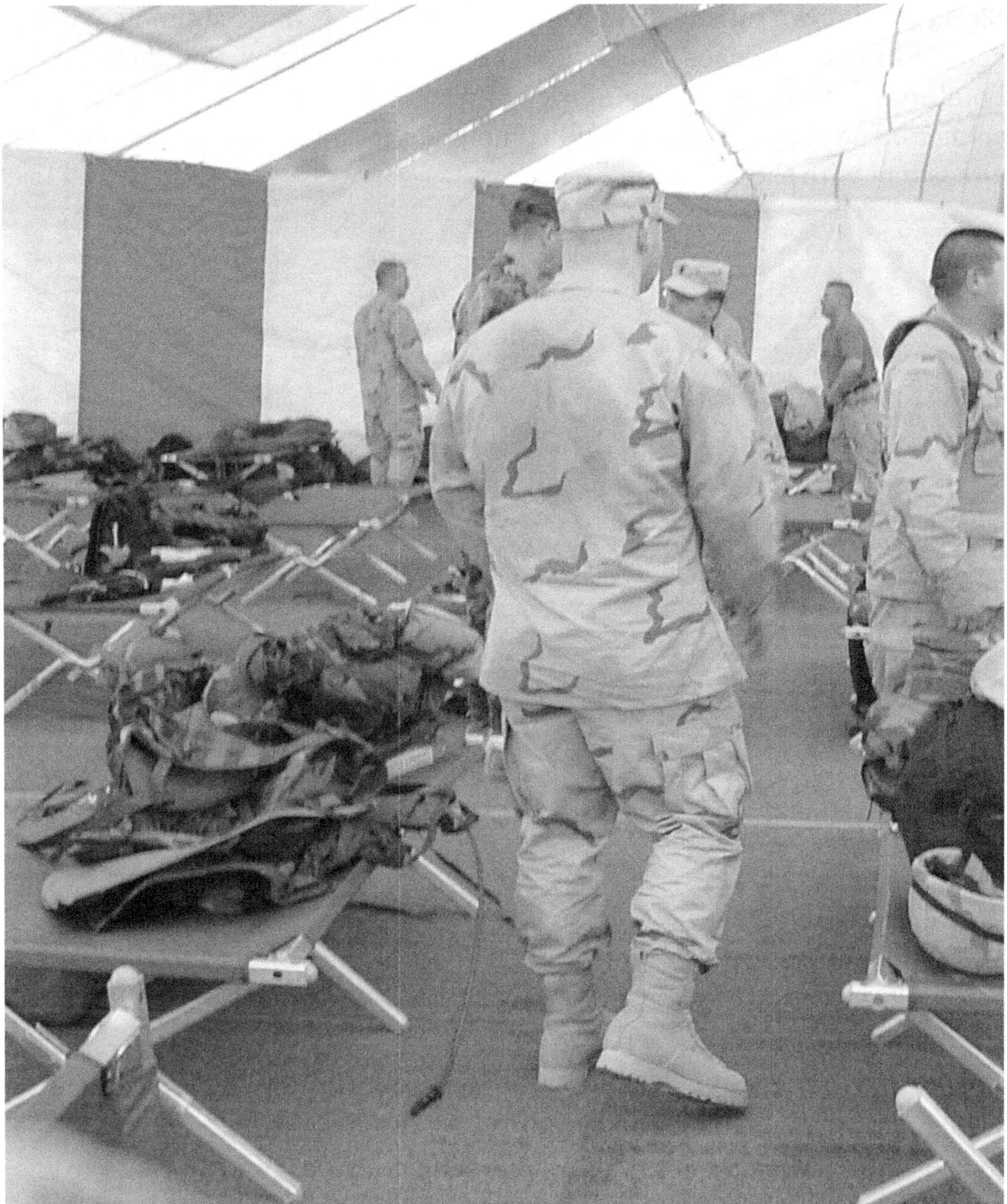

Our accommodations while at Fort Irwin, CA.
(You won't find these photos on military recruiting posters!)

Absolutely no privacy. The showers, bathing, and latrines were almost a mile away from the tents. Picture 1,000 soldiers sleeping under the same roof, snoring, etc. It was awful. However, when you hit the rack, you are so tired from the day's training that you do not care. By the way, where did

Slacker bed down during this month in hell? He, of course, found an air-conditioned office space somewhere on post. He also stayed at the base hotel with his wife at the beginning of training. Not me; I stayed with the soldiers. The chaplain and I buddied up together and remained in the circus tent with the soldiers.

Our soldiers took a long walk to the chow hall for a meal while at Fort Irwin.

Our brigade participated in Fort Irwin's first mission rehearsal exercise (MRE). Big Army (military jargon for the active-duty Army) even built an Iraqi town on the post and had real Arabic-speaking actors role-playing during our training. This situation was quite realistic. Our training was entirely in the field and lasted approximately one month.

The Big Army certifies the training that enables the unit to deploy to combat and engage in battle. Hence, I was unable to access a computer to generate journal entries.

The biggest and worst issue imaginable occurred during this phase of training. LTC Slacker and CSM Hunts lost their individual weapons, 9mm pistols. When these weapons were reported missing, the entire post of Ft. Irwin was shut down. Each and every unit on the post conducted a shakedown inspection. This procedure is protocol in the military, also known as a lockdown. The *Los Angeles Paper* had a field day with this situation, and the lost weapons were never recovered. The battalion's three-day pass was revoked due to the commander's negligence, and morale was at an all-time low. Slacker was facing charges of incompetence, and the brigade prepared for his termination from command. Finally, we get a break.

Not so fast.

To make a long story short, since Big Army had just validated the brigade and its battalions for combat, there was a choice for the powers that be to make. Fire Slacker, as is entirely appropriate, and replace him with another lieutenant colonel or the next in the succession of command, which would be me. However, there is a "catch-22" since the Army validated us with Slacker as the commander; if he were to be replaced, we would have to be recertified with a different commander. Hell, we were days away from shipping out to the Middle East.

So, Slacker gets away with an Army mortal sin: losing his weapon.

Slacker is like a cat; this son-of-a-bitch must have nine lives.

Slacker stays. God help us!

7 Mar 04

I went to bed around 11:30 p.m. and was woken up at 1:00 a.m. by Lieutenant Fernandez, who informed us that we had to be at the TOC brigade at 2:00 a.m. to create the order for our pass. I guess the brigade had second thoughts about punishing the entire battalion for their leader's faults. As I slowly woke up, I thought, "Creating an OPORD at this hour? Really?" Well, we did, and boy, it was a long night. From 2:00 a.m. to 4:00 a.m., we executed the MDMP process and cranked out an operations order for our battalion's pass.

8 Mar 04

This was one of the longest days of my life. I spent the morning briefing our order for Iraq and the rest of the day rehearsing with the brigade. We did a rock walk (rehearsing maneuver exercises — see glossary for further definition) of our movement into the theater.

Later, I jumped on a bus to our home armory in Southern California. I had only one hour of sleep last night.

This is the reality of war. This is one of those small things civilians don't understand. This is something soldiers who haven't been under fire will never understand. As the battalion executive officer, I often had to get only one or two hours of sleep on any given night due to mission necessity. But you will never hear me complain. When you are in combat mode and in charge of soldiers, your adrenaline kicks in, and you do it.

9 Mar 04

On the first day of the pass, I spent most of the day on teleconferences, coordinating the governor's visit to NTC. My poor family watched me work at our kitchen table, with no downtime. I felt terrible for them. They didn't get any quality time with me. I felt guilty.

10 Mar 04

The second day of the pass.

I worked on bus issues. Again, my family watched as I fielded numerous telephone calls. There is not much quality time here, either. I really feel guilty about this situation, especially for my poor family.

I continued working on coordinating buses for our soldiers as I arrived at our home armory at 3:00 p.m. Only three other soldiers got on the bus for Fort Irwin. This is not a good sign.

11 Mar 04

The last day of our three-day pass.

I thought I would say farewell to my sister and her family. Unfortunately, Sherry blew up and shouted endlessly at me over the phone. She asked me, "Why did you have to get a lawyer?" I don't fully understand the big deal with that unless she has something to hide. Annette also got into it with her. Later, Sherry spoke for a very long time with Nina. We had an extended family talk following the calls.

Poor Little Junior was affected by seeing us all upset. It was an unfortunate way to spend my last night at home with my family. Thanks, Sherry Termagants. Tensions were extraordinarily high for all of us, and you just made things worse. Here's the memory that stands out: My last recollection of being with my family in our wonderful home before I went off to war was of my sister yelling at me on the phone at the top of her lungs. I can only pray for her and hope she gets some counseling for her anger issues.

It has been said that mental illness can be pretty difficult for other family members.

12 Mar 04

A tough day. Annette drove me back to NTC early; we left around 7:00 a.m.

We, of course, hit Starbucks on the way up Southern California's Cajon Pass. Governor Schwarzenegger visited today; oh, f'ing joy!

I was worried about our returning troops, or lack thereof. The battalion commander called me over the three-day pass and instructed me to inform the battalion to board the bus one day before the pass expired. I complied with the communication; however, knowing human nature, I was highly concerned with the outcome. Can you believe LTC Slacker wanted to cut his soldiers' passes short by a day? Incomprehensible! I was concerned that the BC would hold me accountable for the noncompliance. I was so concerned. When I showed up at around 9:30 a.m., there were only a handful of returnees. By noon, the deadline, few had checked in. It looked ugly.

Surprise, the governor showed up about 1½ hours late; it was hot and sunny. But he was in rare form, telling stories about weightlifting and motion pictures. I must say he is a superb public speaker.

This was quite a morale boost for our soldiers. I worked at the press booth, overseeing interviews with soldiers. I also interviewed with the *AP* and local newspapers, and Channel 4 interviewed me for their 5:00 p.m. telecasts.

I had a nice send-off with Annette.

Again, the poor girl was in tears.

I am afraid that tomorrow, I will be the one in tears.

13 Mar 04

Today was one of those days often referred to as a pivotal moment in one's life. It ranked up there with the high school prom, your wedding, the births of your children, and the death of a loved one, such as your parents. Today, I finally had to say goodbye to my loving wife, Annette, and daughter, Nina. It was a painful experience for us all. Junior stayed home with a babysitter because we thought it would be too much for the little guy.

I started the day by getting up at 5:00 a.m. and showering; I did not want to smell like I had just come out of a field exercise for my loved ones. I hit the chow hall around 5:45 a.m. to find out the hours had changed; they were now opening at 6:30 a.m. (banker's hours for the Army). I was pissed at the inconvenience because I had planned a 7:00 a.m. staff call at the parade grounds to rehearse today's going-away ceremony.

Our whole timeline was blown to crap.

We got there around 8:30 a.m., and that was when I first saw the ceremony's script. Talk about cramming. I memorized the commands and movements and then rehearsed with the primary staff, company commanders, guidon bearers, color guard, and the Army band.

The event went exceptionally well. I was the commander of troops (COT) and gave the battalion all its commands during the ceremony. I was the eye of the storm, and I was proud to stand before our 1,000 soldiers, especially with my wife and daughter just a few feet away, watching. This event further quantifies my essential role in our organization, and my family could witness this firsthand.

Thousands of family members and fellow soldiers were present to witness the ceremony and bid us farewell as we departed for war. I have never felt so proud. Everyone there saw who was giving out the orders and directing the ceremony, and that was me. I can only wonder if the powers-that-be in the National Guard will remember this moment in years to come. We will see, I suppose.

Following the ceremony, I engaged in the appropriate schmoozing with some of the general officers — sometimes, you have to play the game. Annette, Nina, and I walked to the break area, where we had cold drinks and cookies.

We attempted to sit alone, even if it meant sitting in the blazing sun. We all communicated our love for one another. Annette also told me they saw me on the 5:00 p.m. news yesterday.

I expressed my wishes to Nina so she could treat her brother with kindness and love and be patient with him. I also told Annette how I wished she and Nina would get along. Eventually, we walked toward the buses that would take them off to the visitors' parking area. Annette did not want to overstay her welcome. We all hugged each other and expressed our love to one another. I will never forget the last look Annette gave me; it was the smile of a beautiful woman, just like the one I fell in love with over twenty-three years ago. I so love my wife.

14 Mar 04

Today is another day closer to departure. It is getting close. After starting the staff in the morning, I took an on-post bus ride to attend the church service.

It was a pretty emotional experience, and I donned my dark sunglasses to conceal my emotions from the other practitioners. If you are a genuine believer, how can you not get emotional in church during times like these?

I prayed so hard for my family and my soldiers. It was a moment of reflection for me, thinking of my recently deceased parents, my sister and her family, and my other immediate family members and soldiers. I tried chasing down the priest after the service, but he vanished.

I bumped into MAJ Gritty, the brigade historian. He has completed chapter one of his book for the brigade; I did not want to inform him that he is way behind the power curve, as I am already on page ninety-three of my military memoir. What in the hell does this guy do with his time? Heck,

I am running a battalion, all the while documenting this adventure as often as possible.

15 Mar 04

What I will forever remember about today is that I was incredibly moved by an email my wife sent to my nieces. Annette described the events of our departure ceremony in great detail.

She spoke about my position as commander of the troops and that I was the center of the ceremony. Annette even mentioned that I was participating in time-honored traditions that date back to the Napoleonic era. For the first time, I genuinely believed my wife and daughter finally understood the importance of our efforts and the magnitude of my position. I feel relieved they now understand what I am doing and what we are involved in.

Thank you, God, for that.

It is 9:48 p.m. Good night.

16 Mar 04

We are almost there.

Today's big event, for me anyway, was attending confession. The priest blessed rosaries for us and conducted a Mass of combat deployment, focusing on confession and reconciliation with God. I found it uplifting. They also supplied us with booklets containing many pertinent prayers. I examined the area related to my sins and prepared for confession. I did and am proud to have done so. I also prayed for my parents in Heaven, my immediate family, my sister, and our troubled relationship. I focused hard on my prayers for our soldiers. I asked the priest to pray for them and wanted them all to return home in one piece. He said he prays every day for our soldiers.

I cannot stress enough how vital faith comes into play for soldiers going off to war, at least for me, anyway.

Today, we finalized our packing and did a bit of cleanup. We are allowed to carry one A-bag (a large duffel bag), a rucksack (a backpack containing a lot of equipment, including your sleeping bag), and a carry-on bag or laptop.

Getting all your worldly possessions into such a cramped space is challenging. But soldiers seem to have a knack for this sort of thing. I also had my B-bag (a second duffel bag), which did not make it onto a preflight a week or two ago. (I have been too damn busy running the battalion to worry about my stuff.)

Walking out of the church, I looked up and witnessed a magnificent, clear sky through a beautifully blooming tree. As I continued looking with an overwhelming sense of warmth, I asked my parents to look down on my soldiers and protect them in any way they could. It was quite a great way to leave the church. I believe my parents are looking over my shoulder; I miss them terribly.

The calm before the storm.

17 Mar 04

Today is St. Patrick's Day. However, there is no time to celebrate because today is also departure day. Destination: The Middle East.

Today is our last day at Ft. Irwin, CA, and the last day in the USA for a year or possibly ever.

Sadness.

Fear.

Somewhat helpless.

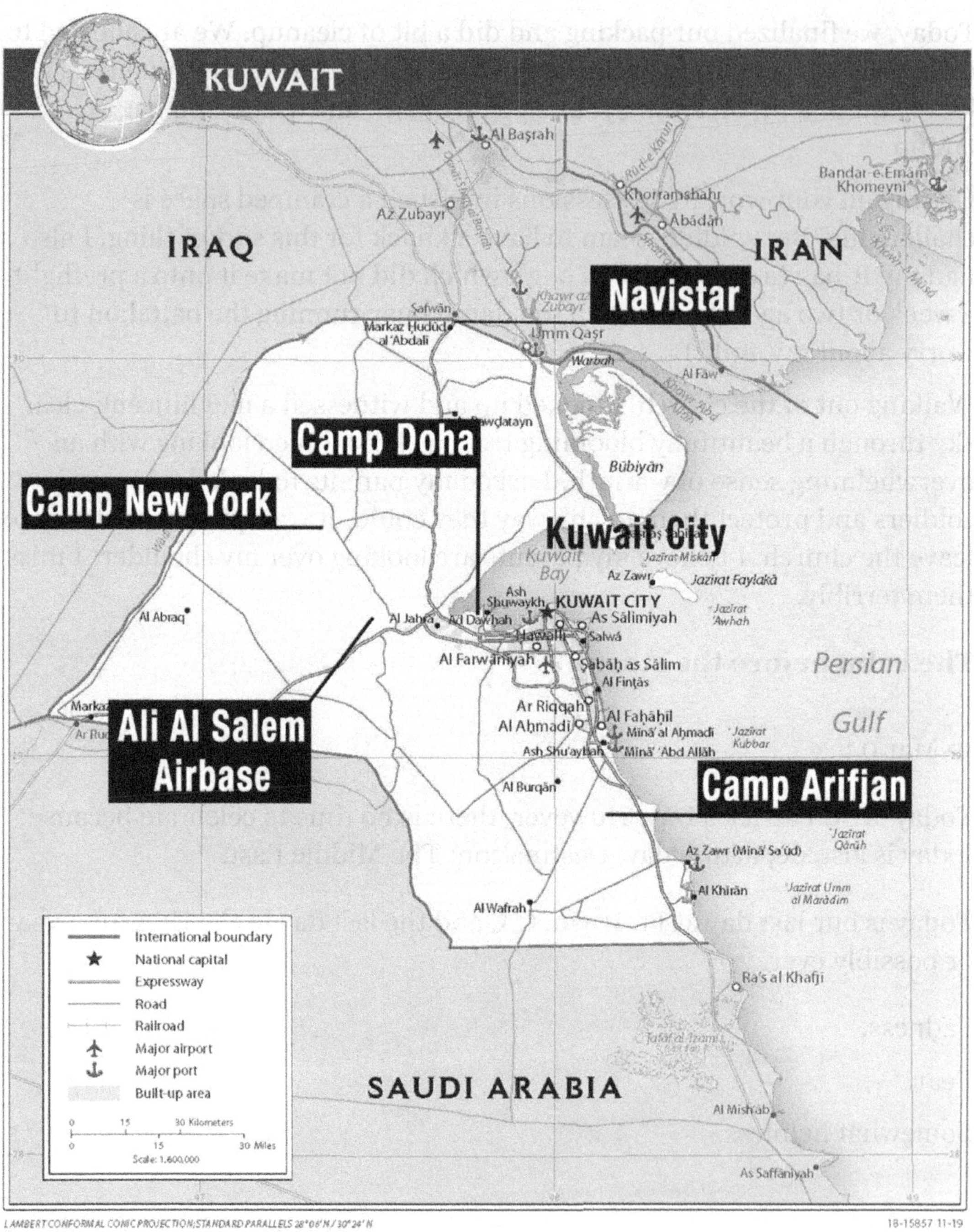

Kuwait is situated south-southeast of Iraq and borders the Persian Gulf.
Kuwait map by John J. McBrearty. Source: The World Factbook, CIA Maps

Chapter 5

Kuwait

17 Mar 04 (continued)

Like many other days, today was no different: rush, rush, rush! I raced to get things done: chow, coffee, and attend the brigade's final meeting here in the States. I had a small number of battalion staff members with me on the second plane. Last night, LTC Slacker left with the main body on the first plane.

We split up our leadership for obvious reasons: If one plane goes down, the other plane with the remaining leadership would be in charge of the battalion. These are the cold, hard realities of war and how our lives now exist.

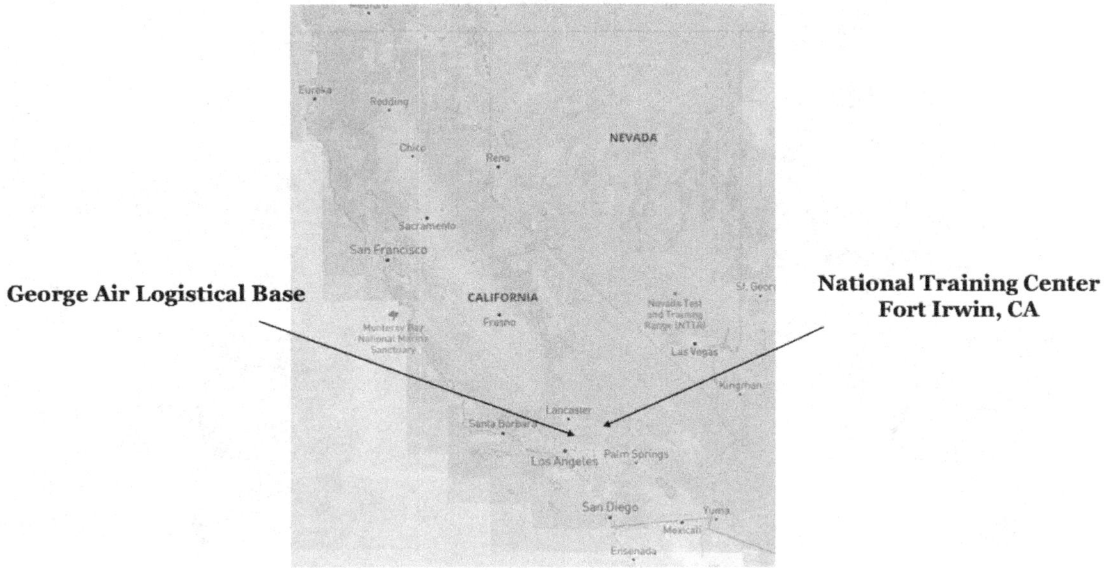

State of California

Los Angeles / Victorville / Barstow area map by John J. McBrearty. Source: © OpenStreetMap (data is available under the Open Database License).

The journey to the Southern California Logistics Airport was like any other Army "goat rope" event: "hurry up and wait," and boy, is that an understatement. We weighed in last night with our equipment and had the flight manifests created. The second flight is on a DC-10; quite a piece of shit for such a long journey. We will leave Victorville around 4:30 p.m., head to an Air National Guard Base in Bangor, Maine, refuel there, and then fly to an Air Force base in Germany. We will change our aircrew there, as it will have been well over twelve hours in the air, not including loading the plane, refueling, etc. The crew was superb, and the flight attendants were gorgeous and friendly.

After venturing on buses from NTC to the logistics airport, we loaded the plane, which took several hours because we had so many weapons and so much equipment.

National Guard Soldiers departing California for Kuwait.

The Veterans of Foreign Wars (VFW) provided a great sendoff. Many members and their spouses handed out goodie bags and thanked us for our service, accompanied by plenty of handshakes and hugs. I recall a gentleman who visited Slacker and me at our armory several months ago when we received the call-up. God bless these VFW members. We will now be qualified to join their venerable ranks.

I have my favorite Linkin Park CD playing on my laptop. If I ever get hit, Annette, Nina, and Junior, listen to and watch Linkin Park's *Live in Texas* DVD to get inside my psyche and emotions. It has provided me with the only source of entertainment for the last four months. I love the band and their music. I have always been a rocker at heart.

So, we did not have Irish Whiskey and green beer on this St. Patty's Day, yet another sacrifice for God and country. Heck, I won't be able to celebrate this holiday for another two years. When I return, I won't even recognize my boy, Junior.

It's all for the price of freedom. I have no regrets. As we made our way to the tarmac, I was overwhelmed by the presence of the VFW folks in the hangar. I wore my dark military shades, which were good as I shed several tears.

SSShhhhh — I am supposed to be tough. Hell, I was too scared to be tough.

Walking to the DC-10, I could not help but take in the picturesque view of the San Bernardino Mountains and experience the beauty of the warm California desert, an indescribable beauty, particularly when you know you won't see it for quite some time, if ever again, and it looks beautiful.

As I approached the plane, I knew these were the last steps I would take on American soil for a very long time or possibly ever. Life gets simple; you understand how precious it is and appreciate all those little things you usually take for granted.

As I approached the stairs, I crossed my heart several times, praying to our Lord. When I got on the plane, I opened the booklet I had received the previous day from the priests; I read the *Nicene Creed*, the *Apostles' Creed*, and several other Catholic prayers. When we started taxiing, I grabbed my rosary and crossed my heart multiple times. I prayed for our soldiers' safe return, my safe return, and for my family.

On the plane, we heard of a hotel bombing in Iraq. Every day for weeks, we have heard of the growing number of U.S. casualties and fatalities, raising our stress and apprehension. Anyone who says they aren't scared is either a fool or lying. I am scared, really fucking scared. I hope that I can hold up and be brave for my men.

God bless the soldiers of the Armor Battalion. May we all return safely.

The stopover in Frankfurt, Germany, was pleasant. A United Service Organizations (USO) volunteer, an elderly gentleman with an Irish brogue, met us. I immediately found Lieutenant McBore, a native of the motherland; the three of us chatted (my father's descendants are from Ireland). They discussed various Irish towns they had visited, with most of the talking done by them. I just took in all that wonderful Irish heritage.

I was able to get online at the USO briefly. I received emails from Annette's brother and his family wishing me well, and from all my cousins on my Hillcrest side. My brother-in-law is a physician, and I always gotten along quite well with him. He is about the most unpretentious doctor you will ever meet.

We are currently flying over Iraq. Some twenty minutes ago, we flew over Baghdad. The captain beside me said, "Look at the lights," as we glanced out the window. It did not look very different from any other big metropolitan area from an altitude of at least 30,000 feet. As we flew south, there were no lights anywhere, just darkness. Eventually, we flew across another city with many lights. There was a sporadic low cloud cover, giving it an ominous flavor. Now that the airplane is descending, we all know what that means: harm's way.

We are here, and there's no turning back now. My heart is pounding out of my chest. I can only think of Little Junior, and I wish to see him again; he needs a dad.

Godspeed.

18-19 Mar 04

I put both days down as I have no clue what day it is.

We are now in Kuwait.

We were delayed at our air point of debarkation (APOD) due to an improvised explosive device (IED) found at the entrance of Camp New York, which was our destination. We were told to bed down, which gave us one hour or so of rack time (military jargon meaning sleep). Up again, chow (the quality wasn't too bad), and off to the buses. I telephoned Annette from the Morale, Welfare & Recreation (MWR) phone system. We had a friendly chat; she said Carmon had sent her and other families an email saying we had made it safely to Kuwait. Annette was relieved, and I am thankful for Carmon's efforts. She had been through it all before when her husband had deployed for Operation Desert Storm.

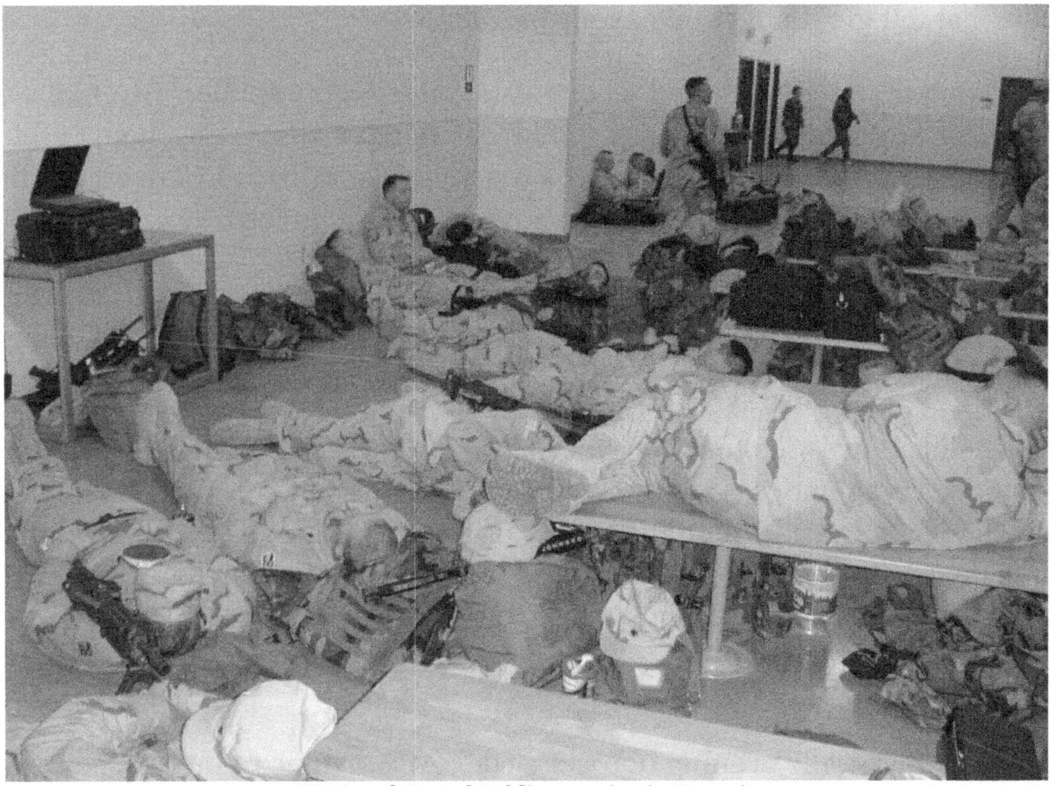

National Guard Soldiers arrive in Kuwait.
(A good soldier can sleep literally anywhere!)

Hopefully, more families like mine will see the valuable assistance that the family support network can provide at times like this.

We drove approximately three hours from the APOD (near Kuwait International Airport) to Camp New York, near Camp Doha, where we will conduct weapons qualifications. We got off the buses, but no one was there to greet us, so we wandered around the various tents, looking for our advance party. I found the HQ tent, and the S3 and BC were sound asleep, like everyone else in the tent. This was our reception? Welcome to the war, I suppose?

I woke them up to report our strength (personnel numbers) and sensitive items (weapons, night observation devices—NODs, etc.). They rolled over and fell back asleep. It turned out they had waited some eight hours to get back into the base last evening due to the IED report. The IED had to be cleared by explosive ordnance disposal (EOD). This is our new reality.

After completing a 20-minute workout of push-ups and sit-ups (which definitely helped release nervous energy), I made my way to the shower. Boy, that felt great after such a long journey.

Following that, I got situated. There were only seventy soldiers in a tent this time, which was still cramped. I hit the rack for a few short hours and then went to chow. LT Ronsen, our current liaison officer (LNO), escorted me to Brigade HQ. We no longer go anywhere without a battle buddy. Again, this is our new reality.

After chow, I returned to the rack and typed on my laptop, creating entries for my journal.

The feeling now is that of apprehension. We were given reports that a company commander, a captain from another unit, and his driver were blown up just outside our gates. It has us thinking. Also, a soldier from the brigade turned a loaded gun on his sergeant. That soldier was taken away for psychiatric evaluation and won't return anytime soon.

The nightmare has begun.

Here in Kuwait, the evening hours are cool. At a distance, you can see an amber glowing bonfire burning. The fire emits a scent of burnt wood in the air. Just a few miles to our north is the border with Iraq. It is there, just waiting for us. Sometimes, I experience overwhelming apprehension, and I must learn to control myself.

So far, so good. I hope I can hold up.

The calm before the storm.

20 Mar 04

It was a rough day for me. The word we got from the general was that all the brigade soldiers coming into theater would spend the first 24 hours doing nothing, just relaxing and sleeping. The reason behind this act of generosity was to get over jet lag. This is a good concept that will prevent injuries, accidents, and other issues caused by fatigue.

As I was cashing in on this concept, at about 9:00 a.m., LTC Slacker approached my rack and said that the staff was running amok and that the tactical operations center (TOC) wasn't set up or tracking anything.

I guess Slacker didn't get the general's guidance.

After flying for 24 hours, I managed to get around six hours of sleep. Oh, well, off I went for what was probably the longest deployment day for me; so much for relaxing, sleeping, and recuperation.

I had a staff call and gave the respective sections specific marching orders, including work priorities and timelines. I instructed Sergeant Major Westminster to get the tracking charts up and the TOC squared away. The TOC was never actually set up that day, and it should be friggen' Army standard operating procedures to do so. There will be more to follow on this subject tomorrow.

21 Mar 04

I was up early, SSS (military jargon for shit, shower, and shave), and then hit the chow hall.

We all eat a hearty breakfast here, as it is very good chow (food). At 8:00 a.m., I conducted a command and staff meeting at the camp's mayor's briefing tent.

At about 8:20 a.m., LTC Slacker busted into the meeting; he was pissed. I called the staff to attention; there was a significant pause as Slacker entered, so I called at ease. Slacker yelled, "I did not fucken' say at ease!" He shouted at me, stating that I did not inform him of the meeting, and he did not like that. I tried to say that I did not want to wake him up and that it was for the primary staff, which was my responsibility, but it was pointless. When he is pissed, look out.

Later, I talked a bit with LTC Slacker; he plans on conducting non-judicial punishment, also known as Article 15s, for the three lieutenants on the advance party for how they so irresponsibly handled sensitive items and generally screwed things up. There was very little planning and coordination before we got here.

When we woke up today, the temperature outside was cool, and the wind kicked up the sand to the point of no visibility. These were harsh conditions.

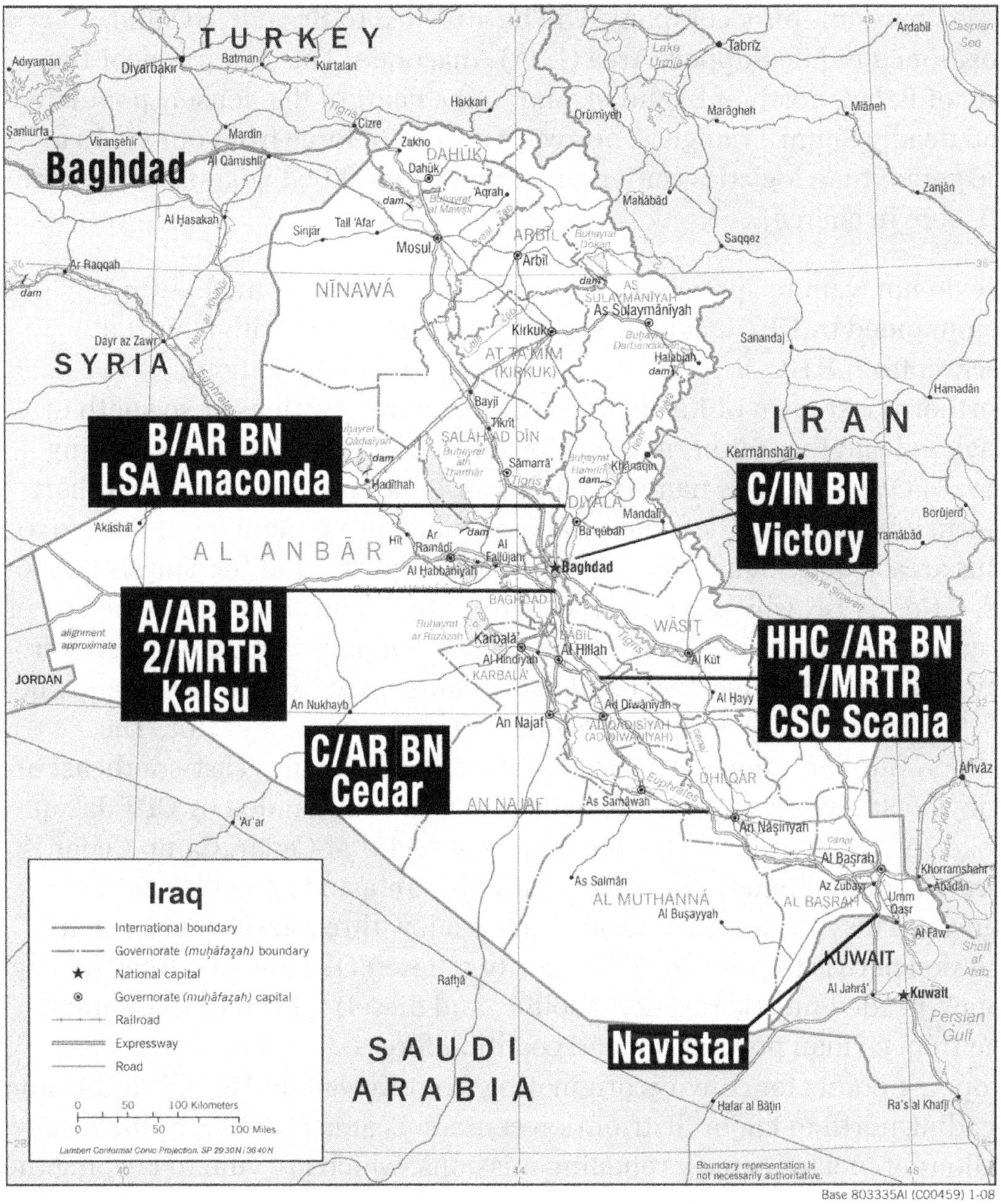

Iraq country map by John J. McBrearty.
Source: The World Factbook, CIA Maps.

23 Mar 04

We held a command-and-staff meeting at 8:00 a.m. at the TOC. LTC Slacker addressed changes to our mission and task organization. The BC mentioned putting me in charge of the Scania area, with him up north with our main effort.

B Co., our tank-pure company, will be attached to Brigade HQ and positioned in Life Support Area (LSA) Anaconda. This camp is near the City of Balad, north of Baghdad, and in the heart of the densely populated and deadly Sunni Triangle. They want a show of force from our tanks at the gargantuan logistics center and air base (frankly, I would do the same if I were in charge).

We do not want to spread ourselves too thin with personnel. A Co., commanded by CPT Crader and 1SG Rangston, along with a mortar section from HHC, is our main effort and is positioned at our northernmost town of Kalsu, approximately twenty miles or so south of Baghdad and the southern tip of the Sunni triangle. Forward Operating Base (FOB) Kalsu was named in honor of First Lieutenant James Robert Kalsu, a Buffalo Bills defensive lineman killed in Vietnam on 21 July 1970. This area is quite dangerous. This is Sunni country, and the Sunnis hate us. C Co. IN BN will also be in the heart of the Sunni Triangle in the city of Baghdad at Camp Victory. However, we do not have tactical control over this unit. Our headquarters company, commanded by CPT Castellano and 1SG Randerson, will cover the Scania sector, our middle sector. The Convoy Support Center (CSC) Scania is located 25 miles east-southeast of Babylon and poses a moderate danger. C Co., commanded by CPT Jason Yoko and 1SG Bruno, covers our southern sector at Cedar. Camp Cedar is located approximately 187 miles south of Scania and 150 miles north of Kuwait. This area is agricultural and has a low threat level. Our primary mission here is to provide FOB security. These FOBs are vital to providing much-needed fuel (kerosene, gasoline, and diesel) for the Iraqi military, the Iraqi civilian population, and coalition forces. Cedar and Scania are stopping points for convoys originating from Kuwait and Saudi Arabia and heading north to major distribution centers. Scania provides millions of gallons of fuel for convoy refueling missions, which are vital to keeping the logistical pipelines open in Iraq.

The Armor Battalion will conduct patrols, guard tower operations, cordon-and-search operations, raids, and numerous vehicle checkpoints to maintain proper force protection of these forward operating bases. The Main Supply Route (MSR) Tampa is vital to keeping convoys safe so they can keep moving to their respective destinations. Relationships with local tribal and civic leaders are also essential for keeping peace in the region.

23 March 04 (continued)

The temperature is rising, accompanied by beautiful, breathtaking sunrises and pleasant sunsets.

Once again, **the calm before the storm.**

Today, we received another report of some soldiers getting hit by IEDs, mainly in Baghdad and Camp Anaconda. We get these reports from the intelligence officers seemingly every day. This has become our new reality.

24 Mar 04

Today was a day of stress.

I got my ass chewed by LTC Slacker in front of my entire staff. He was wrong to do so. We had a heated discussion about selecting our LNO to be assigned to the brigade headquarters. I was so pissed at being ridiculed in a public forum that I just got up from the table and walked out of the tent. Rather than arguing with him in front of others, I made my statement by walking out on him. That said it all. Later, upon my return, Slacker immediately approached me and said that the point I was trying to make was correct and that it was a superb idea, but that I had not communicated it appropriately. WTF?

With Slacker, there is always a twist like that. I was trying to make the case for CPT Ryker, an experienced captain, to serve as our LNO and logistician, based up north in LSA Anaconda, where the brigade headquarters will be located. Also, Anaconda will have the forward support battalion (FSB) positioned there, providing logistical support for us. It would be prudent to have a representative on the ground to monitor our needs and deliver our supplies.

Slacker said he wanted Second Lieutenant Ronsen to fill the LNO position. This would be a terrible mistake as he lacks the experience and logistical knowledge necessary to be effective. During the argument, Slacker yelled, "No, I said no!" I tried to discuss the matter further using a different approach, but he replied, "I said no!" again. Further discussion is attempted, and Slacker flies off the handle: "I fucking said no; what do you not understand about that? No means no!"

I was frustrated.

I attempted to communicate with him about the changes in the mission and task organization. He made the decisions about the placement of every staff member and even every NCO without consulting me or engaging in staff discussion or dialogue. I tried to tell him that I run the staff; they are my responsibility for success or failure, and we all know that is my job. Furthermore, I would like to have, no, I insist on having some input regarding the battalion staff. My attempt to communicate fell on deaf ears. What also bothers me is that none of the decisions were staffed. We learned through the many months of training with the active-duty Army that you always staff big decisions. The only element of change to that concept is time. In a time-restricted environment, you modify the staffing process, but you still staff big decisions. This is Army SOP. We have been trained to conduct the military decision-making process for big decisions, and frankly, we are damn good at it, too.

When encountering a significant change to the plan, we have been trained to attack these issues as a staff and as a team. Of course, we begin the process with the commander's intent and guidance, and sometimes a commander will vary their level of participation in the process. We are not accustomed to having Slacker involved in the process. This time, there was no process at all. There was no discussion whatsoever, and he made all the decisions. This is his right as a commander, but I have problems with the situation as a whole.

Eventually, CPT Ryker was assigned as our LNO, as I recommended. I think Slacker just likes to raise a big stink and show everyone who is the boss. Perhaps this behavior was due to his deep-down insecurities and antisocial personality. All I know is that when you are right, you are right, and I was right in selecting this vitally important LNO position.

I am tired. It is almost 11 p.m., and I am not in bed yet. Tomorrow, we have a 5:00 a.m. wake-up because we will go to the range to zero the weapon, which adjusts it so a target is in its sights.

25-26 Mar 04

Jumping through hoops, the battalion completed various details (military terminology for projects) from guarding posts to moving vehicles from the port to Camp New York. There was not much time for anything. On the evening of the 26th, I worked until 1:00 a.m. because I had to assist the company commanders with preparing their slides for tomorrow's briefing to the commanding general.

27 Mar 04

Today was the most difficult day of my deployment, but it wasn't because of military issues; it had to do with my sister. I received an email from my lawyer informing me that she alleged that my sister quite possibly could have been up to no good without my knowledge. *(No need to elaborate on the details in this forum.)* I felt like I had the wind kicked out of me. It is no wonder my sister has treated me so poorly; I feel she could possibly be guilty of something.

Last night and today, SGM Westminster and I put together a briefing for Brigadier General Killman. The briefing went down flawlessly. Following the general, the brigade command sergeant major, CSM Carr, spoke to us. He mentioned how the 82nd Airborne soldiers in our area of operations (AO) had eight killed in action (KIA), 20 wounded in action (WIA), and numerous accidental injuries in three months. Three of the KIAs were in our Kalsu location; boy, the reality was setting in. CSM Carr spoke of the need to order body bags for the processing of our dead and to have a plan on how to deal with the next-of-kin notifications in the States. He mentioned that these are the facts, and we must deal with these issues because they will happen.

My fear is, will it be me?

28 Mar 04

Today was a pretty good day, possibly because last night's briefing went so well; in fact, we were the buzz of the brigade, and the battalion that had to brief tonight asked to borrow our slide and briefing format.

Oscar Wilde once wrote, "Imitation is the sincerest form of flattery." We are on our game.

Today, I held a meeting for the staff and later a second meeting for the entire command. Our concerns included the safety of our soldiers, equipment accountability, issues with the weapons range, logistical shortfalls, and maintenance nightmares, among others. There are only a few days to go, and there is so much equipment to prepare and load out.

I had distanced myself from LTC Slacker since our incident when I walked out on him the other day. I made my statement to him and everyone there that day, and I think they have come to respect me and the way I do business. I try to operate under the philosophy of reasonability. You must also know when to pick your fights or take a stance. If you complain about everyone and every little thing, you may be perceived as difficult to work with.

Where is the soldier's head right now?

Well, no sex, no booze, no phone, almost no email, presently no exercise, just hard work and jumping through hoops. If you spend much time dwelling on these issues, you can create depression for yourself. I guess I really am too busy to worry about such things right now. In the Army, they say that NTC is a harder gig than actual combat. I am not sure if that is true in our case, but the road to combat has been a difficult one for the brigade, the battalion, and certainly for me. Secondly, the bullets are not real at Fort Irwin, California, but they are quite real here in the Middle East.

We take our personal weapons everywhere we go. I take my M16A2 rifle and 9mm pistol with me to the shower, to chow, and even to bed. I place the M16 under my bed mat, between me and my cot, so no one can see it or touch it without going through me.

My 9mm is at the head of my cot, next to my small field pillow. Again, if someone wants this pistol, they must wake me up to get it.

29 Mar 04

Today went by like a flash.

At 10:00 a.m., we did the back brief and rock drill (an operational rehearsal) with the brigade. It went well.

We received word today that our unit was attacked upon the brigade's arrival at Camp Anaconda, about thirty miles north of Baghdad, the country's capital. Fortunately, there were no casualties.

We got hit by a sandstorm from hell today. The wind increased to around 40 mph, resulting in a blinding dust storm. You could not see more than thirty feet in front of you, and walking in this mess was quite difficult.

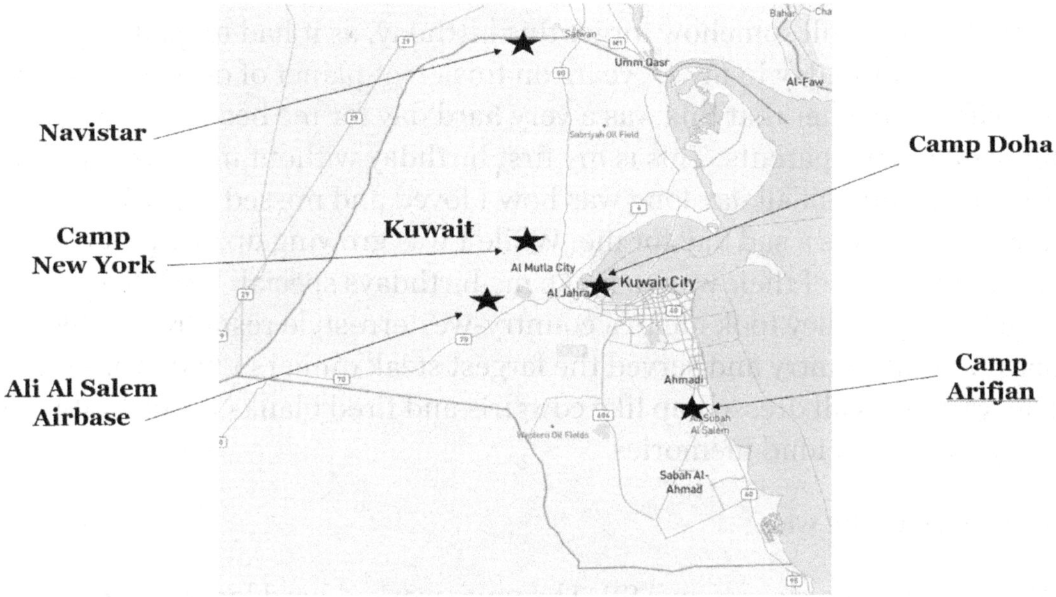

Kuwait Country Map

Kuwait country map by John J. McBrearty.
Source: © OpenStreetMap (data is available under the Open Database License).

We have only a few more days here, and then it is off to a post called Navistar, on the border of Kuwait and Iraq. All units go through there before crossing the berm (an artificial ridge or embankment) as they enter Iraq. From there, we will convoy north on MSR Tampa up to Camp Cedar. We will drop off our Company C there and take the rest of the battalion farther up the MSR to Scania.

At Scania, we will bed down for the night. My serial, Headquarters Company, will stay there, and the main effort (ME) will proceed to FOB Kalsu the following day. Kalsu is a hot zone and dangerous. We are concentrating most of our heavy forces there: A Co., which is at 100%

strength, a mortar section, a scout section, and a forward command cell. We will have a ten-day period to study how our counterparts conducted business. This process is called a "right-seat ride." Once that is completed, we will conduct the official transfer of authority (TOA), the long-awaited day when 82nd Airborne personnel are relieved of their duties in Iraq.

30 Mar 04

Happy Birthday!

Yeah, I think I will somehow forget this birthday, as it had to qualify as the most uneventful day in my 44 years on this great planet of ours. I wrote to my wife, telling her that this was a very hard day for me because all I could think of was my parents. This is my first birthday without my parents alive. All I could think of all day long was how I loved and missed my parents terribly. This was a sad day for me. While I was growing up, my parents always went out of their way to make my birthdays special. I will never forget the time they took me to a country-western-style restaurant. This place was all country and served the largest steak dinners I had ever seen. The waitresses all dressed up like cowgirls and fired blanks to celebrate my birthday — such fond memories.

Now, back to the war.

We are getting so close to our SP. The units worked hard loading their trucks and handling odds and ends. We sent some 21 HMMWVs (Humvees) to Camp Doha to get radio mounts installed.

Earlier, the Brigade COS informed me that they were taking one of our soldiers, SPC Lamas, who is apparently the only graves registration (mortuary affairs) military occupational specialty (MOS) soldier in the brigade. Lamas is also a mortician in his civilian life. I told the COS that Slacker would never go for this and that he would want something in return.

Throughout the day, I walked through the battalion area and observed our equipment being loaded.

Following the rehearsal, I pulled the eight serial commanders aside. As the convoy commander, I told them I needed to know they knew what they were doing. I had LT Ronsen write up a test of about twenty-five questions on convoy operations in this operational environment. I told them I would give them the written test in 24 hours and that I expected them to get 100%. I would not allow them to be serial commanders if they failed to score 100% on this test, because our soldiers' safety was at stake.

All officers agreed to study the materials. Hopefully, it will pay off.

31 Mar 04

I am so darn busy and exhausted that my head is spinning. Every time a fire is put out, another one ignites.

1 Apr 04

I was so tired last night that I did not have the energy to type anything. Hell, I am so tired tonight that I can barely remember anything about yesterday.

Today, I observed my serial commanders conducting their rehearsals and OPORD briefings to their respective units.

Our Washington brothers/sisters cease to amaze me with their lack of consideration for our California soldiers. The brigade staff has snubbed our battalion once again. We were the last unit to arrive in theater and the last to leave for our FOBs. This means that all other units in the brigade had priority for equipment and up-arming (enhanced protection and weaponry) their Humvees. Here we are, a day and a half out from our SP, and we do not get a fraction of our Humvees up-armored as did the rest of the brigade. My Humvee had a vinyl top and doors, with absolutely no armament for protection.

It also irks me that many of our vehicles aren't up-armored, and they could have been. If anyone dies because of that, I will make an even bigger stink about the situation.

I believe my body is physically breaking down due to the excessive workload, long hours, and stress.

Back to LINKIN PARK.

Good night.

2-3 April 04

Today is the long-awaited movement to Navistar, Kuwait.

Armor Battalion soldiers on the move to Navistar, Kuwait.

3 Apr 04

Today is the day we leave Camp New York, marking another milestone in this life-or-death race.

We will convoy to Navistar, Kuwait, approximately one mile south of the Iraq border. It is the final staging area for American troops heading into Iraq.

Our convoy consisted of eight serials, each comprising approximately 30 to 40 vehicles, for a total of 204 vehicles. All vehicles were filled with occupants in every seat, as they represented firepower. The convoy included 600 soldiers, and I was concerned about their safe passage into Iraq.

I guess we aren't your weekend warriors anymore. One weekend a month and two weeks in the summer? My ass!

I instructed all serial commanders to study the Combined Joint Task Force, Operation Iraqi Freedom (CJTF-7 OIF) smart cards and administered a written test on the procedures. The commander in theater developed these battle drills in response to the current threats, including IEDs and ambushes. All commanders passed with flying colors. I now observed the convoy serial commanders conducting rehearsals with their troops with greater confidence, though some did so more thoroughly than others.

Leaving for Navistar was a unique experience because it was the first time venturing outside the wire, north. North is where the enemy is and where hundreds of American soldiers are getting killed. Venturing out of Camp New York, Serial 2, CPT Castellano, took a wrong turn. We arrived at Navistar an hour ahead of Serial 2 (we were Serial 3). The morning was rushed, and I even dressed quickly.

We made it to Navistar just in time to get dinner. Several of the serials had some vehicles go down for maintenance. Self-recovery was the primary means of vehicle extrication, as we were all short on wreckers (recovery trucks similar to civilian tow trucks but much larger). Pulling into the camp, the serial commanders checked into the movement control office to receive a briefing on the camp's layout, dos and don'ts, and intelligence information. The vehicles get staged in a vast staging area, "ducks in a row, and assholes to belly buttons." No living arrangements are set up; you sleep on or beside your vehicle.

There was a porta-potty (portable toilets with no plumbing) for every 100 soldiers, which was tough for women soldiers. We later discovered that the mess hall was the best in the theater of operation, so I was determined to get a bite to eat. Although I wasn't hungry, I ventured the two miles on foot to the mess hall — yes, two miles. If you were not hungry before you left for chow, the walk certainly made you hungry.

The facilities were outstanding: a pizza shop, a Subway restaurant stand, an internet café, a workout room, a basketball court, and a PX. This was the life. Unfortunately, we enjoyed no amenities, just dinner and back to business.

When it came to getting some shuteye (sleep), I improvised in the middle of nowhere, with no tents or cots. I pulled out one of my "Hadjie" (also pronounced "Hodjie," a slang term used by soldiers meaning an Iraqi person) mats, which I had used for the floor around my rack back at Camp New York. I then had the idea of using my sleeping bag cover as a sleeping bag. The cover is a Gore-Tex outer liner that protects the bag from wet weather and keeps you warm. I changed my undergarments and was good to go for a few hours of sleep. It always fascinates me how resourceful the American soldier can be.

Our wake-up call was about 2:30 a.m.; the first serial was to leave at 3:00 a.m. The movement control people kick out these convoys early in the morning because threat levels are lower at that time. They say Hadjie comes out when the weather is good, neither too early nor too late, and not in the heat of the day.

Tomorrow, we cross the berm and enter Iraq.

This is the final **calm before the storm**.

Tomorrow, the storm begins.

Armor Battalion soldiers are preparing equipment for movement into Iraq.

Conclusion

In **From Home to Hell**: *Combat Journal* 1 (Abridged Editon), we witnessed the citizen-soldiers of the Armor Battalion's call-up, mobilization, pre-combat training, certification, and eventual deployment for Operation Iraqi Freedom. This was the largest deployment of California Army National Guard ground-combat troops since the Korean War. As you can see, the business of war is not necessarily glamorous, but it is a necessary evil for the United States of America to survive.

In **First Day, First Firefight:** *Combat Journal 2* (Abridged Edition) discover what happens to the Armor Battalion after they cross the berm into Iraq.

Will the citizen-soldiers be up to the task at hand: combat action?

John J. McBrearty

Adapt, create, and move forward.

FIRST DAY
FIRST FIREFIGHT
Combat Journal 2
John J. McBrearty
Lieutenant Colonel, U.S. Army (Retired)

What is next?

FIRST DAY, FIRST FIREFIGHT

Combat Journal 2

(Abridged Edition)

We left off in **FROM HOME TO HELL:** *Combat Journal 1* (Abridged Edition), with the Armor Battalion in Navistar, Kuwait, awaiting orders to cross the berm into Iraq. After many months of pre-combat training and certification, we will finally see whether the citizen-soldiers of California's Army National Guard are ready for the task at hand: combat operations. This time, it is no drill; it is not training, and the bullets are real. The responsibility of commanding troops in combat weighs heavily on the shoulders of the battalion executive officer, Major Darby Hillcrest. Find out whether Major Hillcrest is up to the challenge in Lieutenant Colonel John J. McBrearty's second volume of the *Combat Journal* series.

Glossary of Military Terms

AGR	Active Guard Reserve members are National Guardsmen/women who work full-time supporting their respective units. They receive the same benefits as active-duty Army personnel and are eligible for an active-duty retirement.
AT	Annual training is a two-week program, typically conducted during the summer months, for reserve component military personnel.
BIAP	Baghdad International Airport — formerly Saddam International Airport, also known as Camp Victory.
Green Zone	It is located near central Baghdad, heavily guarded, and home to several former presidential palaces. U.S., coalition, and Iraqi authorities live and work here, which is a dangerous location during the war.
Gung Ho	A slang term meaning esprit de corps or enthusiasm.
Hilla	Hilla (also spelled Hillah) is a major city in central Iraq, located some 62 miles south of Baghdad along the Euphrates River. Because of its proximity to the river, it is predominantly agricultural.
HMMWV (Humvees)	High Mobility Multipurpose Wheeled Vehicle (Colloquial: Humvee).
Hooah	Heard, Understood, and Acknowledged, originally HUA — a slang term representing esprit de corps.
In Country	Military jargon for being in Iraq or other foreign countries in combat.
In Shaa Allah, inshallah	"God willing" in the Arabic language. This phrase is used by the Iraqis quite frequently.
Joe	A term of endearment for enlisted soldiers. It's also slang for coffee.
Kalsu (FOB Kalsu)	FOB Kalsu is located approximately twenty miles south of Baghdad and on the southern tip of the Sunni Triangle. This is an extremely dangerous area.
MDMP	Military Decision-Making Process. This is the U.S. Army's 7-step analytical process for making calculated decisions in tactical and garrison matters.

NGB	The National Guard Bureau serves as the headquarters for all National Guard forces.
NOD	A NOD, or NODs, is a night observation device.
PX	A post exchange is similar to a shopping center on a military installation.
Rock Drill or Rock Walk	Setting up terrain models on the ground, literally using rocks and other natural items to resemble actual terrain. It is an effective tool for commanders to visualize the battlefield accurately. Units use this method to rehearse maneuver exercises by placing various units on the terrain model and going through the mission's phases.
S1, S2, S3, S4, etc.	These are primary staff positions at the battalion and brigade levels, including personnel, intelligence, operations, logistics, and others. For the remainder of the positions, refer to the references cited below.
Scania (CSC Scania)	CSC Scania is a convoy support center that does what the name implies: it provides fuel, food, water, and other support to American service personnel, coalition forces, the Iraqi military, and select Iraqi civilians passing through Iraq's main north-south transportation corridor (MSR-Main Supply Route-Tampa). It is located 25 miles east/southeast of the City of Babylon and approximately 90 miles south of Baghdad. This camp has moderate danger.
Squared Away	Military slang meaning "having things in order."
Stovepipe Leadership	A stovepipe commander does not allow information to flow up, down, or within their organization, hindering communication and cooperation with others—narrow-minded thinking.
Tallil (Tallil Airfield)	The Tallil Airfield is located adjacent to Camp Cedar. The airfield has a low threat level for coalition forces.
Termagants	A violent, intimidating, controlling, argumentative, loud shrew of a woman. A turbulent bully with a bad temper. (William Shakespeare)
Theater	Theater, or in theater, refers to the theater of operations, where combat operations take place.

TOC	The TOC, or tactical operations center, is typically the central hub for all planning and operational matters.
TSD	Training support detachments are composed of active-duty Army advisors who enhance the training of reserve component units. They are considered experts in their respective fields.
UCMJ	The Uniform Code of Military Justice is the military's legal system.
VBIED	A VBIED is a vehicle-borne improvised explosive device, also referred to as a car bomb.
Victory, Camp Victory	Camp Victory is situated in the Green Zone, located in the heart of the Sunni Triangle in Baghdad, Iraq. It occupies the surrounding area 3 miles east of Baghdad International Airport (BIAP). This is a dangerous location.
XO	The XO or BXO is the executive officer, the second-in-command of a unit such as a company, battalion, or brigade. ***(The dude who gets things done!)***

References:

FM 25-100 (Field Manual-Training the Force)
FM 7-0 (Field Manual-Training the Force)
JP 1-02 (Joint Publication-DOD Dictionary of Military and Associated Terms)
FM 6-0 (Field Manual-Command and Staff Organization and Operations)
FM 22-100 (Field Manual-Army Leadership)
FM 6-22 (Field Manual-Developing Leaders)
ADP 5-0 (Army Doctrine Publication-The Operations Process)
AR 600-8-22 (Army Regulation-Military Awards)
AR 735-5 (Army Regulation-Property Accountability Policies)
Department of Defense Dictionary of Military and Associated Terms.
https://www.supremecourt.gov/opinions/URLs_Cited/OT2021/21A477/21A477-1.pdf

Lieutenant Colonel John J. McBrearty's brain (the primary source of this information).

Sergeant Major Westminster is the operations sergeant major and is the second-highest-ranking enlisted soldier in the battalion. He is built like a former football player, which lends itself to being a bit intimidating. Cynical and cocky would best describe this man. Every other word out of his mouth is a complaint about how "f'__d up" things are.

Sergeant First Class Dwellers is a worthless excuse for a soldier. He is afraid of his own shadow and won't leave the wire (encampment) for anything short of a direct order. Most of the battalion hates this guy's guts. A real "buddy fucker!" (Sorry for the language, but get ready for a bumpy ride on the following pages with this guy.) We later discovered that Dwellers was a snitch for the inadequate battalion commander. A real prick.

Wives:

Annette Hillcrest is the wife of the battalion executive officer, MAJ Darby Hillcrest. She is a 40-year-old attractive schoolteacher and former ballet dancer. This mobilization could not have come at a worse time. This woman is a saint and holds the family together back home.

Carmon Dwellers is the battalion family support coordinator and the wife of SFC Dwellers, the battalion's personnel NCO. Carmon is a no-nonsense type, but it's anybody's guess how she ended up with a loser like SFC Dwellers; she is a nice person.

Saving the best for last.

Sherry Termagants is the older sister of Major Hillcrest. As demonstrated by her actions, Sherry disapproves of our country's efforts in the global war on terrorism. She certainly does not support her brother and his participation in the war in Iraq and has cut off all ties with his family. The demise of their relationship weighs heavily on Major Hillcrest and his family. However, you cannot rationalize with the irrational. Which Hollywood actress comes to mind who resembles Sherry? Glenn Close's character in *Fatal Attraction*. Scary!

Michael Termagants is Sherry's husband and brother-in-law of Major Hillcrest. Michael and Sherry are two peas in a pod. He is an enabler for his unscrupulous wife. Michael is like a white-collar gangster you see in the movies. But this guy is not in a movie. He is real-life evil.

Acknowledgments

As in many successful endeavors in life, we often didn't get there by ourselves. We had a helping hand along the way. I want to thank the following individuals for giving me that helping hand.

To my parents and in-laws, to my wife of four decades, my daughter, my son, and my grandsons, to my close friends Dave, Duke, Chris, and Raf, to the Gross family and all the other friends and relatives who wrote to us and sent my unit greatly appreciated care packages while we were in combat, to Callaway Golf, to *Stars & Stripes* newspaper, and to the other news agencies keenly interested in our unit and who covered the deployment.

To *Veterans' Voices*---Amy Johnson, *Sports Central*---Jim Hill, Chris Hayre, Glenn Shimada, and Gordon Allen at CBS Los Angeles for your support of our veterans.

To the U.S. Naval Sea Cadet Corps, which gave me my start in uniform when I was just a teen. Thank you, Lieutenant Commander Stolle and family.

To Sa Bum Nim (Master Karate Instructor, 7th degree black belt), Jung Hung Kim, my Taekwondo instructor in my formative years. The man who instilled in his students respect, discipline, and courage.

To the United States Marine Corps, which gave me my foundation in leadership.

I am grateful to Valley Forge Military Academy & College for providing me with the academic foundation to succeed in life.

Thank you to the Army National Guard for allowing me to serve my country, state, community, and soldiers.

To all my previous commanders and NCOs who mentored and taught me along the way. They trained me to be firm but also fair with my soldiers.

The Department of Veterans Affairs, Loma Linda, CA, and all their excellent, hardworking, dedicated staff.

My literary mentors include the following:

John Cole, a U.S. Marine Corps veteran of the Vietnam War, is the author of six novels and a book of poetry, *Giants and Friends*, honoring his fellow Vietnam combat Marines. John is a brilliant writer, a good friend, and a mentor. *Semper Fi, Brother!*

Written by Veterans (WBV) creative writing group, California State University, San Bernardino. WBV authors include Cindy Rinne, Pete Schreiber, Fred "Bo" Dunning, and others. They are all published authors and quite an eclectic bunch of patriots.

Andreas Kossak the book's editor, mentor extraordinaire, and founder of WBV. He is a former adjunct professor at California State University, San Bernardino, CA, and a member of the prestigious Writers' Guild of America. This man is truly inspirational, as his selfless service to our veterans is unparalleled.

To my Lord and Savior.

God bless you all.

John J. McBrearty

About the Author

Lieutenant Colonel John J. McBrearty, USA (Retired), served honorably in our nation's military for thirty-two years in positions of increasing responsibility. He has the distinction of serving in both the U.S. Marine Corps and the U.S. Army. His career included combat duty and spanned the globe, with operations in Iraq, Kuwait, Japan, Australia, Mexico, and Thailand.

John J. McBrearty, a member of Phi Beta Kappa, is a magna cum laude graduate of the Valley Forge Military Academy and College, as well as Temple University. He also holds a master's degree in American history from the American Public University. He served as an Assistant Professor of Military Science at California State University, San Bernardino, and Claremont McKenna College. His unique multi-service career, increasingly significant military assignments, professorial experience, and combat action give him keen insight into the importance of American history and freedom.

John J. McBrearty's other works have been published or featured in *Sports Illustrated, Golf Week, Golf Digest, Stars and Stripes, Callaway Magazine, Grizzly Magazine, Los Angeles Times, Orange County Register, Main Line Times, News of Delaware County, Daily Bulletin, San Bernardino Sun, Press-Enterprise, Veterans Breakfast Club Magazine, Military Writers Society of America Dispatches Magazine,* and others.

Having joined the ranks of the "retired," John is active in several Veterans Affairs support groups. Known as "COL Mack," he is also a member of several veterans' writing support groups, where he serves as a literary mentor and advisor to fellow veterans. He volunteers as a golf instructor for junior golfers and fellow veterans and enjoys a friendly round of golf from time to time.

John has published over twenty books across multiple genres, including American history, military memoir, leadership, self-help for veterans, how-to guides on writing and getting published, golf instruction, and children's books.

John was featured on segments of CBS Los Angeles's *Veterans' Voices* (American History books) and *Sports Central* (Children's Golf books).

John has a large social media following that includes dozens of literary YouTube videos garnering thousands of views and followers.

John won 1st Place in the National Veterans Creative Arts Festival for creative writing, which earned him a Veterans Therapeutic Arts Scholarship.

Upon his retirement from the military, John was recognized for excellence by President Barack Obama.

Books by John J. McBrearty

Leadership Books

LEADERSHIP LESSONS (1), *From Operation Iraqi Freedom*
ARMY VALUES (2), *Lessons in Leadership*
OIF LEADERSHIP 101 (3)

American History

AMERICAN HISTORY, A VETERAN'S PERSPECTIVE (1), *Essays, Reflections, and Reviews*
COMBAT ESSAYS (2), *Operation Iraqi Freedom*
MY MILITARY LIFE (3), *From Sea Cadet to Lieutenant Colonel, U.S. Army, NO REGRETS!*

Military Memoir Book Series

FROM HOME TO HELL: *Combat Journal 1* (Abridged Edition)
FIRST DAY, FIRST FIREFIGHT: *Combat Journal 2* (Abridged Edition)
ON PATROL IN IRAQ: *Combat Journal 3* (Abridged Edition)
THE BURDEN OF COMMAND: *Combat Journal 4* (Abridged Edition)
WHISPERS OF MUTINY: *Combat Journal 5* (Abridged Edition)
IRAQ TANK BATTLE: *Combat Journal 6* (Bonus Edition)

CITIZEN SOLDIER, *Operation Iraqi Freedom: A Memoir*

How to Write and Get Published Books

LITERARY WORKS OF JOHN J. MCBREARTY
HOW TO GET PUBLISHED ON AMAZON
Volume 1, *A Guide for Beginning Writers*
Volume 2, *From Publishing Zero to Publishing Hero!*
Volume 3, *Inspirational Quotations and Other Writings by a Combat Veteran*
Volume 4, *YouTube 101: for Writers Only!*
Volume 5, *How to Write a Book Review.*

Sports Books

HOW TO GOLF (1), *Children's Golf, How to Teach the Golf Swing in Four Easy Steps*
HOW TO GOLF (2), *Driving Range 101*
HOW TO GOLF (3), *Golf Course 101*
HOW TO GOLF (4), *Advanced Golf Techniques Anyone Can Learn*

Children's Books

CLAIRE THE MAGICAL DOG (1), *Claire Meets Dean and Jacob*
CLAIRE THE MAGICAL DOG (2), *Claire Introduces Baby to Golf*
CLAIRE THE MAGICAL DOG (3), *Claire Digs to China*
CLAIRE THE MAGICAL DOG (4), *Claire Celebrates Veterans Day*
CLAIRE THE MAGICAL DOG (5), *Claire and the Service Dog*

Motivational Books

Secrets for a Successful Life Book Series

VETERANS' SURVIVAL GUIDE (1), *PTSD Is Not a Four-Letter Word!*
MILITARY SCHOOL (2), *A Foundation for Success*
HOW TO WIN AT LIFE (3), *A Combat Veteran Shares His Secrets*

Anthologies

ANTHOLOGY 5, *Written by Veterans*
MESSAGES FROM THE BACKROOM, *V.A., Loma Linda, CA*

See JOHN J. McBREARTY

Please leave your feedback, thoughts, and reviews on Amazon.

See Lt. Colonel McBrearty on CBS Los Angeles *Veterans' Voices.*

See Lt. Colonel McBrearty on CBS Los Angeles *Sports Central.*

Photo courtesy C.B.S., Los Angeles.

Follow John J. McBrearty's other literary works on social media:

Legacy

"Every generation is tested. Ours answered the call."

John J. McBrearty
Operation Iraqi Freedom
2003-2005

Veterans
Crisis Line

Military
Crisis Line

1-800-273-8255 PRESS 1

Veterans
Crisis Line

NEW NUMBER,
SAME SUPPORT.

Dial 988 then Press 1.
Share it with your networks.

Made in the USA
Monee, IL
07 November 2026

67980867R00096